MIGUEL ANGEL RODRIGUEZ SEGURA

The Secrets of KNX: The Ultimate Beginner's Guide to Master Home and Building Automation

Original Title: "Los Secretos de KNX: Como Dominar la Automatización de Viviendas y Edificios con KNX "

If you need any information related with this book, please contact me on instagram @startupiando.

Second edition

ISBN: 9798648218154

Translation by Julieth Milena Sanchez Jimenez
Translation by Miguel Angel Rodriguez Segura

This book was professionally typeset on Reedsy.
Find out more at reedsy.com

Contents

Why should you read this book?

First congratulations on choosing KNX for your professional training. It is a great protocol that has created a milestone in the automation of homes and buildings, it is also made up of large companies and the KNX association has professionals of incredible technical and personal level. They put their hearts to this and the truth is that they do a phenomenal job. This book is not intended to replace conventional training, it is needed and I encourage to take the KNX courses, but this book is a must-have supplement if you want to be successful in this area.

As a consultant and auditor in automation projects I have seen and solved several problems, many of them would have been easier to prevent than to solve. If you are a KNX partner you will agree with me that the vast majority of the information in those long books of your courses have not been of much use to you in real life and even much of what you needed was not really there, even after to have invested good money and time in them. You will agree with me that real life projects are much more complicated than the practices you did in your demo bags and that the controlled environment of the classroom is far from a real project where unforeseen events suddenly appear and mistakes are part of the process. If you already have more time in KNX you will agree with me that a simple advice at the right time would have saved you a lot of money and time in the projects, you will agree with me that maybe the theory they gave you needed to be landed and refined to be able to be used efficiently, and it is not the fault of your tutors since to reach that level you need to be very good and have a lot of experience only that the formality must be met, there are things that escape because there is no space for them and others that we simply have to dictate. I firmly believe that we are in this world to contribute to others from what we do best and this is what I do best, after having gone all the way formally in KNX, after having dedicated years of study in undergraduate and four postgraduate courses, after Having done projects and met many people, this is what I do best and this is my way of contributing to the development of society. I am convinced that regardless of your level in KNX, this book will give you valuable information that will help you to be successful in KNX, it will help you to get more clients, improve your projects and project your career correctly. After finishing this book and putting into practice what is described here, your way of thinking will completely change and the benefits will be immediate.

Here you will not find ostentatious formulas or purely academic concepts, you have had enough of that in college and I am sure you agree with me that the vast majority have been of little practical use in real life. Possibly in a project if you are going to need to dust off some formula but you can easily find them all on the internet if you know how to look for them. The really important thing is to have clear concepts and a complete vision of the subject, with that you will be able to easily identify what formula you need, in order to search for it and easily find it.

This book compiles the most relevant theory of all KNX courses and the one that is important but not found in them, that is, what you will really need for your KNX projects. Additionally, it has condensed all the practical information that escapes in most of the courses. It also brings together the mistakes of many people so you don't make them and the knowledge developed by others over the years so that you can incorporate it, thus taking a shortcut to excellence. Here you will also learn how to create technology, not just to be a consumer of it like the vast majority of people, so if someday you want to be a manufacturer this is the best way you can take to start opening your mind. Finally, this book gives you a projection of the sector of home and building automation that will allow you to be at the forefront for a few more years.

With the purchase of this book you can request me technical material for free.

Enjoy it and any comment, question or feedback will be very welcome, you can find me on instagram as **@startupiando.**

Cheers and let the game begin!

I

BACKGROUND

This first part of the book is very important, whether you are an engineer or not, because it will give us a basic context before going to see KNX as such. Before we even talk about KNX we have to agree on a couple of concepts such as domotics, inmotics, open protocols and closed protocols.

"There is a delicate balance between honoring the past and losing oneself in it" Ekhart Tolle

1

Home and building automation (Domotics and Inmotics)

Human beings have evolved in order to better satisfy each of their basic needs, and that is why, over time and technological advances, requirements are solved more efficiently. From the earliest times, man has insatiably sought a better quality of life and well-being for himself and his close social groups, making use of what he has at hand at the time; necessity that has driven the developments and discoveries to apply them in these fields.

Around the 70s, a large part of what we know today as domotics begins, a term that is not so popular today in the American continent, however, it is growing progressively. Home automation begins with the research and appearance of the first automation devices based on the still existing technology: X-10. With all these new advances, the search for more ideal buildings, homes and generally constructions was awakened, with which a great evolution is also seen in the architectural field that complements these wishes for more technologically advanced spaces.

Around the 80s and 90s, these technologies and automatisms began to be incorporated, especially in office environments and corporate buildings, an advance that was highly favored with the annexation of structured cabling systems (SCE) at the end of the same decade, at the time from which the concept of intelligent buildings was adapted for all this (An intelligent building was considered at that time, one that had in its SCE structure, which highly benefited the transmission of data, voice, control of some electronic devices and security .). Consecutively these advances are preparing to apply in more residential structures, which brought with it new needs in equipment development, and adopting for this time the term of home automation that is specifically related to the application of automation science and technology within a house; while for automation topics with a focus on tertiary sector facilities: office buildings, hotels, universities, housing facilities in buildings, airports, large stores, etc. The term of French origin applies more: Immotic.

In avant-garde issues, for current homes it is not plausible that they lack basic services, therefore the future of homes is aimed at ensuring that all homes are domotic at least in an elementary way, which will guide automation issues in home automation and immotic as a basic need for homes and buildings in general, just as today are light, gas, water, etc. However one of the main drawbacks that home automation has encountered is that very few people are willing to invest in high costs; which fortunately is a topic that is currently more versatile due to the increase in manufacturers and developers of these technologies, as new solutions are added to new needs.

In general, home and building automation is the integration of technology in the different subsystems of a building, whether residential, commercial or tertiary. The interdisciplinary work of engineering, computing, telecommunications and architecture has pushed the automation of homes and buildings to the level of sophistication that they have today. Due to the relationship of these multiple disciplines, we are faced with new concepts associated with domotics, immotics, smart buildings, bioclimatic buildings, digital homes, among others.

Domotics and its development in favor of the creation of smart homes are complemented by the growing advances in the fields of telecommunications technologies aimed at the treatment of information and communications. The interaction between home automation and information management and communication technologies leads us to the concept of digital homes, which, although they are heading in the same direction and complement each other, carry different notions; specifically, domotics is exclusively associated with the technical management and control of home automation, while the digital home refers to the teleservices offered thanks to the advancement of telecommunications; where the is the subject of entertainment and remote connections. Therefore, it is most appropriate to mention that digital homes include the concepts and applications of the home automation. However, in general, the services of the digital part, control, communication, entertainment and data processing are contained in the description of domotics (a home with a broadband connection: ADSL, fiber optic cable, etc.); making the issues of teleservices and telecommunications relate to the same term of domotics.

In modern buildings it is common to find different systems focused on the management of the electronic equipment located there, normally in charge of controlling audio, video, security, HVAC, lighting and blinds, among others; However, the delight and need to implement these new technologies is increasing in architectural works that are even integrated into buildings of yesteryear and that handle classic designs, but that include highly technological functionalities in their environments. The objective of these systems is to optimize their specific operation and increase the quality of life of the occupants of the building. The different demands of users, added to the increasing presence of applied information sciences, communication at home and attention to collective needs in many cases exceed traditional facilities, which is why solutions that worked independently were created. , such as sensors or even video intercoms, among others, and home automation seeks to integrate all these systems within the same network so that their supervision and control are more effective and simple. All these systems are always considered within the generality of the pillars of domotics, and within which you can find a wide range of activities and equipment that can be integrated into a domotics and immotics installation.

With the imminent advancement of technology and the massification of the Internet of Things, together with the digitization of information, the number of electronic systems and devices in buildings has increased. This makes of vital importance the integration of the same in a single environment and the consequent interaction between them. Both domotics and immotics have as a fundamental column the integration of different building systems with the aim of integrating different functions and tasks. Allowing their interaction in the same integrated home automation network for energy efficiency, safety, comfort and autonomy. Our homes are increasingly digital, in which technologies and devices that receive, transmit, store and process information converge.

With all the evolution that has been presented in the associated topics, the one associated with home automation protocols stands out in the market. The different protocols for home and building automation can be divided into two main classes: Closed protocols and open protocols. Each of them with their respective advantages and disadvantages according to the particular application. Being these protocols in particular those that allow communication between different physical equipment is possible if they are part of the same network.

The protocols are procedures that directly intervene in the communication between all the home automation devices and the way of controlling them. With the passage of time different communication protocols have been developed, some specifically to be applied in the home automation and immotics area and others adapted to be used in the same areas.

2

Open protocols

O pen protocols are those based on an international standard independent of any particular brand, with common software and open technical specifications for different manufacturers. This in order to achieve full interoperability between different products from different manufacturers in the same installation, with full functionality.

Open protocols are developed by consortia, organizations or associations. Among the most relevant are:

• **Zigbee:** It is a radio frequency standard based on IEEE 802.15.4 widely used in the automation of homes and buildings. There are three main types of ZigBee devices that are coordinator, router, and end device. The main difference is that the end device cannot route traffic, the routers can route traffic and the coordinator apart from routing traffic is responsible for forming the network in the first place. The basic architecture consists of a coordinator that manages the network and various devices connected to it. Zigbee can achieve a data transmission rate of 250kbps, although 40kbps is usually sufficient for most applications.

• **Lonworks:** LonWorks was created by Echelon Corp in 1988. It is an open standard for home and building automation with more than 300 manufacturers on the market. LonWorks is a peer to peer (P2P) network and as a means of communication you can use twisted pair, Ethernet or even carrier currents through electrical installation. The configuration of the network is done by means of a tool called "binding" regardless of the software of each device. It is a standard technology for organizations such as ASHRAE, IEEE, ANSI and SEMI among others.

• **BacNet:** BACnet was first published in 1995 by the permanent standard project committee SSPC135 of the American Society of Heating, Refrigeration, and Air Conditioning Engineers (ASHRAE). The current standards for BACnet are ANSI / ASHRAE 135: 2004 and ISO 16484-5: 2007. Its architecture consists of four layers which are the physical layer, the data layer, the network layer and the application board of the ISO / OSI model. Among the main advantages of BACnet is its specific design for real estate and that its open protocol nature allows the integration of different products and systems. BACnet is more focused on building management and automation than on field devices. Among its strongest applications are "Building Management Systems" (BMS) and HVAC management.

• **KNX**: KNX is an open protocol that has been on the market for decades and is currently a leader in home and building automation. The KNX Association was founded in 1990 with headquarters in Brussels (Belgium) and was originally called the "EIB Association". This entity had as its objective the promotion of smart home automation and its applications, in general, and the EIB system, created by several manufacturers of recognized prestige, in particular. In 1999 the association merged with two other European associations, specifically BCI (France): it promoted the Batibus system and the European Home Systems Association (Netherlands): it promoted the EHS system. As a result of this unification, the name was changed to that of the "KNX Association". KNX is approved as an international standard ISO / IEC 14543-3, as a European standard CENELEC EN 50090 / CEN EN 13321-1 and as a Chinese standard GB / T 20965. It also handles different means of communication including twisted pair, radio frequency. Ethernet and carrier currents through conventional electrical installation. KNX has a defined system topology, it is made up of areas, lines and line segments. This topology allows projects of up to 57375 different devices to be carried out in the same system, making KNX the most widely used system for home automation and professional imotics, at the field device level. It also has more than 300 different manufacturers with a wide portfolio of products in everything related to home and building automation, these products are programmed through a single software (ETS) regardless of the manufacturer of the particular device.

The main benefit of open protocols is that they are supported by different manufacturers, even allowing different products from different brands to work together with full functionality and without the need for additional hardware or software.

The protocols or open standards provide a wide range of applications available in the building, giving the customer freedom from not being tied to a particular brand or company.

Allowing the easy expansion or updating of the systems and the easy implementation of new products to be developed.

3

Proprietary protocols

Closed or proprietary protocols are those developed by a particular company and whose technical specifications are not in the public domain, only the companies that develop it have direct access to the protocol, users can access it only under payment of licenses. These protocols arise from the development of products by a certain brand usually focused on some area of home automation and imotics. Among the main closed protocols are:

• **Vivimat:** Developed by the Spanish company Dinitel, a pioneer manufacturer in Spain. The range of its vivimat products is governed by a centralized ideology, with the ability to perform basic automation tasks in a residential environment. Thus, all the products of this brand lie in a control unit (depending on some versions), the unit in which the system programming is located, which is responsible for making decisions. Apart from being a centralized system, the installation can grow in a semi-distributed way as it presents a communication bus based on the RS485 standard, through which the system can be expanded. The platform can be configured in all its versions through the screen and adjust timers, scheduled tasks, scenarios, among others. The "vivimat project" software is only required in case the installation includes some screen references. Vivimat emerges as a platform with a vocation to the real estate area, so it is quite comfortable in its prices, with basic programming. In its latest versions they bet on a higher-end market, but maintaining a segment of affordable housing with basic automation functionalities.

• **At-Home**: It arises from the Airzone company, pioneer of the ALTRA group, and its focus is to complement the installations with airzone systems. It is a system that follows a type of distributed architecture, where its equipment and components each have their own intelligence, capabilities to process information and communicate on the same bus. Its most outstanding feature is the user interface, displayed on a monochrome screen with a temperature probe included; allowing you to create interfaces by zones and create controls in the same way for each space.

• **X-10:** Distributed in Spain by the home systems brand, based on carrier currents, all signals are transmitted over the low-voltage network or by radio frequency. It is a distributed system, where each component processes and sends information. The X-10's capacity is very limited, as its design was based on living environments. Manages equipment addressing by codes of device and house codes, for up to 256 active equipment. For remote control, this system uses high-frequency signals, transmitted with a carrier frequency of 120 kHz, a procedure called pulse code modulation (PCM). The inputs are only open for periods of time to receive the emission signals, this with the aim of suppressing the disturbing influences of the network. These fractions of time occur after the grid voltage passes through zero, since it is the moment in which the probability of disturbance or possible failures is less. The orders on the devices are produced through datagrams, which require 880 milliseconds; fact that affects the user's perception, cataloging it as a slow system in its response action. The X-10 system has its own software called "active home" to program and view, which is simple as well as the installation of the components on the network. However, at present it is considered an unreliable system, largely due to multiple communication interferences since it uses the energy network to transmit information between computers.

• **BUSing:** Developed by the Spanish company Ingenium, located in Asturias. It is a system with distributed architecture very similar to the At-home system. BUSing was developed to use its own bus as a transmission medium, but over time, it evolved to implement its protocol over RS485 or TCP / IP networks, and equipment with wireless technology is also offered by its manufacturer. One of the most outstanding characteristics of this system is its ability to obtain very basic function packages at a very low cost, becoming more competitive than vivimat, with its economic version but less powerful.

• **Rako**: Along with the creation of the company in the United Kingdom, this system is also created. It is a proprietary system that uses radio frequency for data transmission and has a distributed architecture. Specifically designed for lighting control, a field in which it currently specializes. For this reason, Rako, is usually not considered as a home automation system, for not integrating most of the functionalities associated with the theme. However, it is original and useful in terms of wireless control issues. It is very simple to implement, and reduces costs on site as long as you only want to control the lighting theme.

• **Lutron**: United States state company founded in the 1950s. It has positioned itself as the largest manufacturer of lighting control in the world. It allows to control on, off, regulation and scenarios with lighting. Although it also manages curtains of its own manufacture and display systems, its focus and main strength is lighting. Its installation is simple and it handles different lines with wireless communication. The system requires a "hub" to manage the devices. They stand out among its main lines Caseta and EcoSystem.

• **Leviton:** United States state company founded in 1906, with a broad portfolio in the residential, commercial and industrial electricity sector. It was one of the first companies not to require a "hub" since its devices use the WiFi network. Use Z-Wave. Its main focus on home and building automation is lighting, however its portfolio also encompasses audio/video, energy efficiency, entertainment and security.

• **Control4:** United States state company founded in 2003, with a broad portfolio for home and building automation. Its main strength is audio and video management, it is widely recognized for entertainment control in automated buildings. However, it also has products for lighting, HVAC, remote and local control, interconnectivity, controllers and security.

• **DIY:** There are a wide variety of wireless devices that can be controlled through hubs. Among them we can highlight brands such as Broadlink, Orvibo, Google Home, Amazon Echo, among others. These systems are focused on reaching the end user directly and for these they are known as "Do It Yourself". They usually communicate with field devices via radio frequency and provide control of the system through a mobile application.

• **Vantage:** United States company acquired by the Legrand group in 2006. Being lighting specialists they have a portfolio of products for home and building automation in lighting, audio, HVAC, security, remote and local control. As well as relay actuators and both binary and analog inputs.

• **Creston:** United States state company founded in 1968 with a broad portfolio in home and building automation. It is a well established global system for automated home and building control with solutions for lighting, audio, video, HVAC, security, blinds and communications.

• **Insteon:** It is a proprietary technology of the Smartlabs company, launched in 2005. This system is focused on lighting and blinds, with complementary products for HVAC and security. It uses a "hub" and can be integrated with DIY systems like Amazon Echo. The system becomes more robust the more devices you have.

• **Luxmate:** It is a system of Austrian origin with distributed architecture, created by the lighting company Zumtobel Licht. Like Rako, it is not considered a home automation system, since it only performs digital lighting and motor control; However, it is a highly accepted system in the tertiary sector. It offers remote connections through the internet, which facilitates the maintenance of the installation. The sensor called Heliometer stands out a lot, which placed on the roof of the building, is capable of controlling and regulating lighting thanks to the capture of solar radiation, the orientation and geographical location of the building. Another important feature of Luxmate is its ability to perform active maintenance, detecting and locating faults in the luminaires. It also has a memory capacity, with which it can store scenarios for subsequent executions; Other systems require special scene modules for this.

II

FUNDAMENTAL THEORY

This second part of the book condenses the most relevant information from KNX courses as well as the theory that escapes from these courses but that you will need in your projects. Theory is always important as long as it is accompanied by practice, so we will start with the fundamental theory necessary to do KNX projects.

"He who likes to practice without theory is like the sailor who sails ships without rudder or compass and never knows where to anchor" Leonardo Da Vinci

4

A little bit of KNX history

The aim of this book is to give you all the theoretical tools and practices so that you are extraordinary in the area. The process of a project begins before the project itself and it is there where this information will be very useful to you, when you are talking to a client for the first time about the subject, when you are giving a conference or even to better position yourself before your colleagues.

The KNX association was founded in Brussels (Belgium) in 1990 and the most experienced will tell you that its original name was "EIB Association". Its main objective was to promote home automation and still automation in general, and of course its own system called EIB, which was also multi-brand.

At the time the EIB association was not the only one working on the subject and for 1999 it moved with the BCI association in France that promoted the Batibus system and with the EHS association in the Netherlands that promoted its namesake system, EHS.

After the merger the name of the EIB association was changed to KNX, as short for Konnex. The KNX association aims to:

• Develop a unique, stable and affordable technology system with the aim of increasing its general acceptance in all markets, and expanding the current market (mainly tertiary buildings) to the residential market.

• Integrate the existing home and building control systems into a common standard that serves as a platform for future developments.

• Define and improve KNX specifications in relation to: Protocol, different transmission media, configuration modes and application specifications.

The fact that KNX is worldwide implies that it has references on all five continents. Several million successful KNX projects can be found not only throughout Europe, but also in the Far East, North and South America, which can also serve to provide customer security and again seek a better position in secrecy. More than 370 member companies in the world offer more than 7,000 groups of KNX certified products in their catalogs, which provides a wide range of application areas.

It is also important to keep in mind that KNX is an open protocol which implies that it is approved as an international standard (ISO / IEC 14543-3), a European standard (CENELEC EN 50090 and CEN EN 13321-1) and as National standard in countries like China (GB / T 20965). The great advantage of this is that it ensures continuity in time for future changes or extensions of the projects, in addition to the interoperability between products from different manufacturers, which is a great advantage … well, if you know how to take advantage of it.

The KNX application areas are mainly:

Well we already have a little context and a perspective of what we can do with KNX. Now we are going to go into a more technical part, the fundamentals of KNX.

5

KNX Fundamentals

With some of the clear history we move on to the fundamentals of the KNX system, this part is important as it can save the life of your project... literally.

The first thing you should be clear about is that KNX is a distributed system, so there is no central controller that handles everything, each team has its own microcontroller with its stored program. This is a great advantage since in case of failure of a particular equipment, which is not very frequent, the operation of the entire system is not compromised since the other equipment continues to operate normally ... of course if you get damaged the lighting actuator, even if the button continues to work, the light will not turn on. However, if your button also activates blinds or multimedia, this will continue to work without problem. Another example, if the equipment that is in charge of generating the visualization is damaged, you can continue controlling the system from the other field devices such as buttons or screens.

This sounds good, right? … But calm, is not all honey on flakes, being distributed also implies that you will have to program each one of the teams and relate them appropriately with their different variables and relationships associated with the themselves. Which you will learn efficiently in this book.

To communicate KNX equipment you have different options, that is, you have different means of data transmission that you can use at your discretion. The transmission media you have available are:

• **Twisted pair:** This is a standardized and certified cable that mainly comes in two formats: two or four wires. This cable connects the equipment physically. Two wires are used for the communication bus, which incidentally also has 30VDC power for the equipment. And the other two wires, which are optional, are used to send additional power to equipment such as touch screens, for example, this happens because being a multi-brand standard, there are technical specifications that the equipment must meet to carry the KNX guarantee and among them is the amount of current that can be consumed through the communication bus, which is an average of 10mA, equipment such as touch screens require a higher current consumption for their operation, so they have to take the remaining power from an additional power supply. One of the most common questions is if you can use a different cable than the certificate for your projects, and the answer is yes. If you can however you have to take into account that the distance and transmission calculations that you will see later are made for certified cable, like the connectors, therefore if you use a different cable the system may work for you but its stability and reliability will be compromised. We are here to do things the best way, so I recommend you always use certified cable.

• **Powerline:** This transmission medium is the same power line that runs through the building and you must have the neutral available. Are there equipment for this transmission medium? … yes. Are they used frequently? … not. The main reason integrators decide to forego this option is because the range of manufacturers' Powerline equipment is very limited. It is also true that the carrier currents can be susceptible to noise and the phases must also be coupled to make good communication, other protocols do it very well.

• **Radio Frequency:** The name is sufficiently explanatory, KNX also offers the option of communicating wirelessly by RF, which is very useful in real life because rewiring or civil engineering to send a KNX cable is not always as easy as it sounds. . With the radio frequency you must bear in mind that it is not paradise either and if you have a significant amount of equipment the electromagnetic noise that is generated can cause interference and a malfunction of the system.

• **IP:** Finally there is the option of Ethernet / Wifi, which allows KNX equipment to communicate through the local internet network or even remotely. For this you need a team that converts you from twisted pair to IP since for now KNX teams as such do not communicate by IP, but require a converter that communicates the twisted pair with the local network. However, sooner or later this will have to change and there will be KNX equipment such as push buttons or actuators that communicate directly to the local network via WIFI.

The normal thing is to use the twisted pair as the main means of communication since it provides robustness and stability, complementing with IP and radio frequency to achieve flexibility and efficiency in the installation, obtaining maximum functionality. KNX equipment can mainly be divided into three large groups, which are:

• **Actuators:** They are all those equipment that generate an action on a specific load. Among them you will find actuators for lighting, for blinds or awnings, for HVAC, for security, for audio and video, among others.

• **Sensors:** They are all those equipment that convert a physical variable into digital information. For example, buttons, screens, motion sensors, lighting, wind, temperature, air quality, among others.

• **System equipment:** They are basically the rest of equipment... those that neither sense variables nor act on the loads but are necessary for the operation of the system. For example power supplies, couplers, gateways, interfaces and protective equipment.

All these equipments are programmed through a unified software independent of the brand, the ETS.

Well we already know what types of equipment exist and what means we have available to connect them. Now the next step is to know how these teams are organized correctly for a project, let's go for it!!.

6

KNX Topology

In KNX there is a defined topology scheme that you must follow if you want your project to work properly. For this there are three fundamental concepts that we must be clear about: Line Segment, Line and Area. Let's tackle each of them to finally see how far our twisted pair KNX installation can grow.

Line and Line Segment: The line is the most basic structure of the KNX topology, it results from connecting a maximum of 64 twisted pair equipment to a KNX power supply as shown in the next figure.

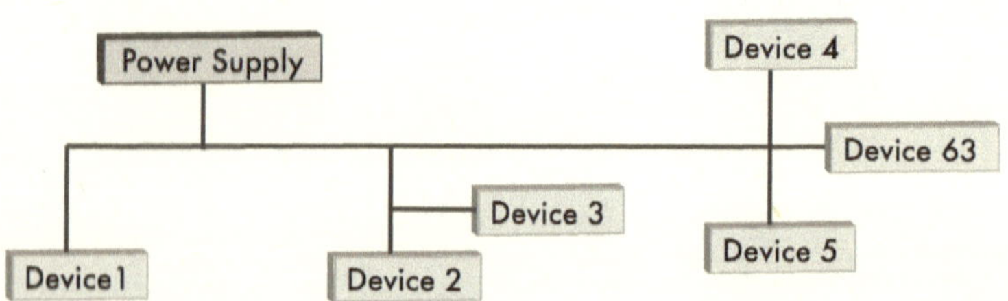

Here it is allowed to connect the equipment in line, star, tree or a combination of the above, but be careful not to make rings or loops.

It should be noted that the amount of equipment you can connect depends on the specific consumption of each equipment and the KNX power supply that you select.

Now if you need to connect more equipment you can start to grow the topology, the next step is to use the line segments. These are extensions that allow you to connect 64 additional devices for each segment, in total you can have 4 line segments as illustrated in the following figure, that is, 256 devices in total.

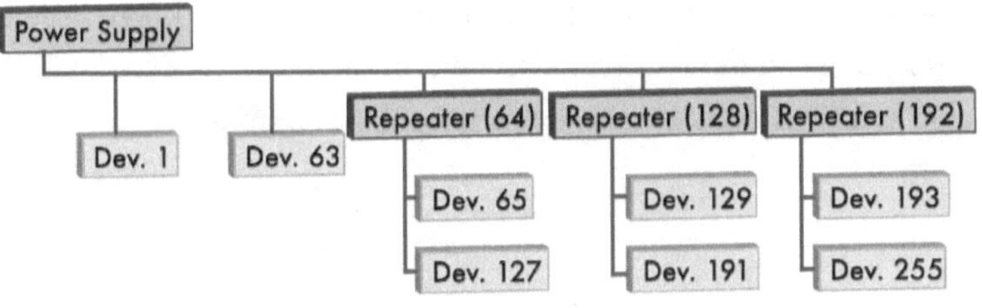

Area: We have already seen that several line segments, maximum 4, come together to form a line. Similarly, lines come together to form one area, up to 15 lines can be contained in one area.

To form an area, a main line is defined, in which up to 64 devices can also be placed, here it is not worth putting repeaters, to this main line the lines that make up the area are coupled with a maximum of 15.

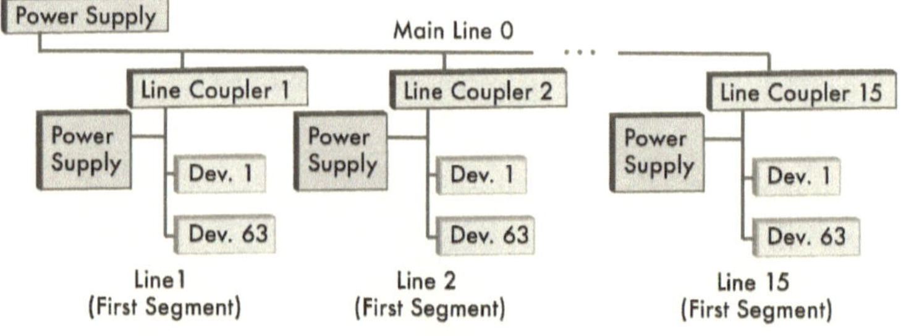

In an area where your lines don't have repeaters, that is, line segments, you can connect up to 1000 twisted pair devices, not bad, right ?.

However, the topology can grow even further by interconnecting different areas through a trunk line or also known as a backbone. 64 devices can also be connected to this trunk line, also without repeaters or additional line segments.

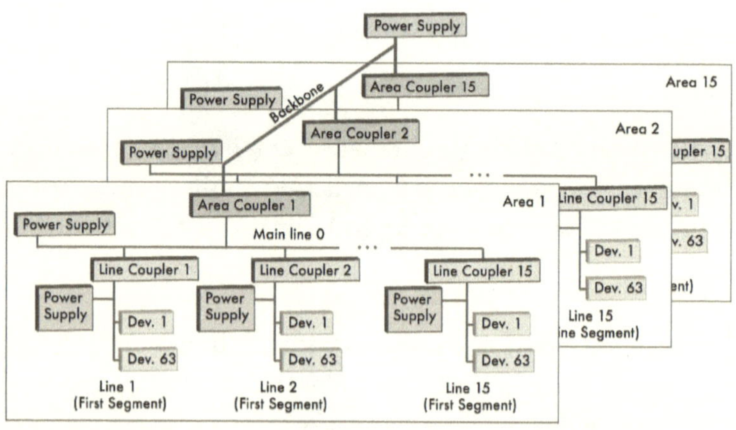

By connecting different areas you can reach up to 15000 bus components, without using line extensions, or 58000 components using line repeaters to add additional segments on the bottom lines.

The decision of the topology does not depend solely on the number of teams but also on their functionality or the design of the particular project.

Until now we know in general where KNX came from, its basic foundations, what types of equipment we can have and how they are organized in the topology of a project. Now we will see how they relate to each other and how they identify.

7

Addressing

Depending on the place in the topology where each of the devices is located, it will have a physical address. This address has the AML format, where the number "A" refers to the area to which the equipment belongs, the number "L" refers to the line where the equipment is located and the number "D" is the number that the device has.

As we saw previously we can have 15 areas, 15 lines and 256 teams per line. Therefore, the physical address of each device will be:

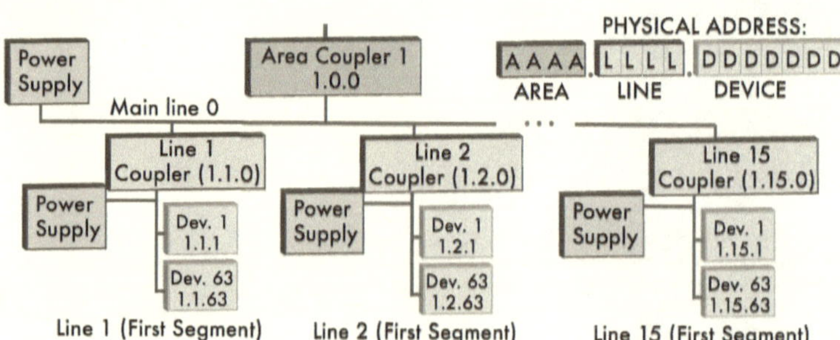

This physical address must be unique within the installation and its main function is to enable communication with the devices mainly from the ETS, to send programming telegrams or modify parameters of the equipment in question.

This address has a fixed 16-bit format as we saw earlier in the image, however the representation in both the programming software and the documentation is decimal.

Usually the devices come from the factory with the physical address 15.15.255 and to change it the programming button is physically pressed on the equipment and the new address is programmed from the programming software, when pressing this button a red led is usually illuminated on the device indicating that it is in programming mode. Although not all computers have a button to literally press, some are done magnetically or digitally.

You may wonder, if the physical address is only to communicate with the equipment from the programming software, how do the computers communicate with each other?...
Great question! To answer it we are going to introduce a new concept: Group address.

Group address is like a room where everyone in there listens to all the messages that are said there, without having to communicate one by one. That is, the messages are sent to a certain group address and all the objects that are related there listen to it. Landing the analogy, each KNX equipment has different objects such as the object to turn on a light in case it is a lighting actuator. If you want to turn on this light, you must associate the object of turning on the actuator light with group address "X" and send the order to turn on the light to group address "X".

We have a maximum of 65535 group addresses to make our associations, taking into account that the group address 0/0/0 is reserved for the transmission of multicast messages, that is, messages that are addressed to all available bus devices.

These group addresses are programmed in the ETS programming software, for which three options are provided:
- 3-level structure (Main / Intermediate / Subgroup)
- 2-level structure (Main / Subgroup)
- Free structure

These structures serve exclusively to achieve a more organized and systematic representation of projects. The most used is the 3 level, however, it is your decision how organize your projects and the important thing is that it be as clear as possible so that a person who sees it in 3 years manages to understand it in a way without having to spend months deciphering what you had in mind. It is important to keep in mind that there is a maximum number of group addresses for each level:
- Main: 31
- Intermediate: 7
- Subgroup: 255

Therefore, it is recommended that you define your own structure to make the implementation of projects and at the end of each project remember to deliver a detailed report to the client along with all the documentation of the project itself.

These group addresses are assigned to each of the communication objects of the KNX devices manually via ETS, if you are working S-Mode or automatically if you are working in E-Mode.

8

Group objects

Now, it is clear to us how each team is identified within the KNX topology and we know that there are group addresses that allow us to make the associations of the different variables so that the teams interact with each other, since those variables are called group objects.

Group objects are the variables associated with each of the KNX equipment. These objects depend on each individual equipment and are enabled depending on the parameters that are configured in the ETS. For example, an On / Off lighting actuator has different group objects even though it apparently only has the function of turning the light on and off. Let's imagine that it is a single channel, that is, it only turns a light on and off. This equipment will then have at least the switching group object that is responsible for turning on the light directly but it will also have a group object that gives the status of the luminaire in question. This status object is an output from the actuator that tells you if the light is on or not, which is very important and we will see it in the practical part but for now let it be clear that they are different associated group objects to the same action, in this case turn on the light. We will also have group objects that depend on the functionality of the equipment, for example timing objects, scenarios and other additional functions.

As you can imagine, just as there are different types of variables, there are also different types of group objects. The shortest being 1 bit and the longest 14 bytes. These data types are standardized, here are the most common:

Code => Data type

1.yyy => Boolean, like switching, move up / down, step.

2.yyy => 2 x boolean, as switching + priority control

3.yyy => boolean + unsigned 3-bit value, as up / down regulation

4.yyy => Character (8-bit)

5.yyy => Unsigned 8-bit value, as dimming value (0..100%), blind position (0..100%)

6.yyy => 2 8-bit complement, as%

7.yyy => Unsigned value 2 x 8-bit sign, as pulse counter

8.yyy => 2 2 x 8-bit complement, as%

9.yyy => 16-bit float, as temperature

10.yyy => Time

11.yyy => Date

12.yyy => Unsigned 4 x 8-bit value

13.yyy => 2 4 x 8-bit complement

14.yyy => Floating 32-bit, as temperature

15.yyy => Control of access

16.yyy => String -> 14 characters (14 x 8-bit)

17.yyy => Scene number

18.yyy => Scene control

19.yyy => Hota + data

20.yyy => Enumeration 8- bits, such as HVAC mode ('auto', 'comfort', 'wait', 'economic', 'protection')

It is important to keep in mind that group addresses only allow objects of the same type to be associated with them. Therefore, if you want to turn on a light from a pushbutton, for example, you should look for the pushbutton control object of the type On / Off 1 bit and the On / Off object of the luminaire to associate them in the same group address. This is very useful since if for example you take the lighting regulation group object and try to associate it with the switching direction, you will get an error, however it is important to be careful because as you saw there are objects that despite having a different nature They are very similar and being of the same size will let you associate them in the same group address, such as the switching object and the switching status object.

The concept is quite simple and really these group objects are only spaces in the memory of each device, we will see a couple of important considerations before going into the details of each subsystem of an automated building.

9

Flags

Let's go back to the analogy of group directions to introduce a new concept.

Thinking of group leadership as a room where there are several people, group objects, everything that is said there will be heard by everyone. But now what if we want to improve communication and define if each one listens to the messages, or even if that particular person can speak? Well for this, flags are used to generate certain permissions when interacting with messages in group addresses.

Remember that the group objects are ultimately only spaces in the memory of the devices, well the first thing is to define if we want these objects to have communication enabled, we can also define if the registry can be modified, if it can be read the information contained therein, whether this object transmits any changes that are made to it or even how it reacts to responses from other objects.

Each group object has 6 flags:

• **Communication (C):** This flag is the main one since it enables or disables the communication of the group object. In other words, if this is deactivated, the object cannot send or receive telegrams.

• **Write (W):** This flag allows writing to the group object. That is, to write new information or overwrite the stored information, this flag must be active, otherwise it will not be possible.

• **Reading (R):** This flag allows the information in the group object to be read by means of a reading telegram. For example, if we want to check the status of a group object, this flag is necessary.

• **Transmission (T):** If this flag is active, the group object in question can transmit changes in its value through the communication bus.

• **Update (A):** This flag in object X indicates if the response message sent by another group object should be taken as a write command in X, that is, if the value of register X should be modified, or if must ignore.

• **Read on Init (I):** This flag indicates that the group object in question will send its value after restarting the communication bus due to a bus voltage failure.

The flags are configured by default in the devices and for the basic use of the same it is not necessary to modify them. However, it is important to be clear about the concept and function of each one for slightly more advanced tasks or even for diagnosis in existing projects.

10

Couplers

When we talked about topology we saw that there were devices called couplers to join different lines or areas, because the time has come to understand what they are and what they are for.

The couplers in a nutshell are a filter, end.

They are in charge of filtering the telegrams that pass through them, that is, if the coupler is working as a line coupler, their function is to define which telegrams leave the line in question and which ones enter it. I mean, it's literally a filter.

And these filters can be configured to allow everything to pass, to allow nothing to pass or to only allow telegrams that go to off-line equipment or that come from offline to equipment online. The immediate question is:

How do I define which telegrams should pass or should enter?

Well, you don't do it explicitly, unless you want to, since when you associate the group objects within the group addresses the ETS itself creates the filter tables for the couplers. Of course you can still modify them manually if you require it.

Let's remember that KNX has different transmission media, therefore there are different types of couplers. The couplers are commonly twisted pair (TP) to twisted pair (TP), however there are also couplers between different transmission media.

For example there is the twisted pair (TP) to Ethernet (IP) coupler, this is widely used since it allows to optimize the physical installation and even improve the communication speed of the system. The function is the same that we already explained simply allows us to join twisted pair lines or areas through the local network, without the need to lay a physical wiring that joins them. These couplers are also known as KNX / IP Routers and through them we can also program the system. Here an important consideration, if we are using an IP coupler below it in the topology, another IP coupler cannot be placed since it would generate conflict in the handling of the addresses, that is, if we are using, for example, an IP area coupler, the couplers of line from that area will need to be wired.

There are also twisted pair (TP) to power line (PL) couplers. This works in the upper part, in the twisted pair installation, as a coupler while in the lower part, in the power line, it works as a repeater for therefore there can be no more repeaters in the line power line.

Finally we have the radio frequency (RF) twisted pair (PL)

couplers. What more than couplers like this, we can see them as repeaters since KNX RF devices are not organized in a specific hierarchy within the topology.

And well, at this moment you will wonder what type of coupler should I use if I want to couple a non-KNX equipment or system to KNX? Well, although the name is directly associated with that, the answer is: To do it, what you need is a Gateway.

11

Gateways

Gateways are devices that are used as gateways or interfaces between two different systems or protocols. In other words, they are a kind of translator that allows two systems that speak different languages to understand each other and to interact with each other.

As we saw in the beginning of the book, there are open and closed protocols for the automation of homes and buildings, each of them has its advantages. There are brands that make proprietary systems with unique devices and that work very well in a particular task, these can be integrated through Gateways in a KNX project. Likewise, there are open protocols with very particular focuses that can also be integrated through Gateways.

In other words, a KNX system can be complemented with different protocols and both proprietary and open systems through Gateways. The concept is clear and it sounds easy enough, but at the same time there is a space, a cognitive limbo that prevents us from practicing. Let's imagine you already have the Gateway, and now what??

The general procedure and concepts are practically the same regardless of the Gateway, if you have it clear to start it, a brief look at the manual of the equipment in question will be enough to refine details. So we will illustrate the procedure and concepts with an example.

Suppose you have a KNX installation in an industrial production plant managing lighting, blinds and air conditioning. But now you want to include the measurement of energy consumption in the system.

The first thing is to check if there are native KNX equipment that meet the technical requirements of the project. Let's assume that the available KNX energy meters do not meet the requirements, the import is too complicated or they are simply out of budget. So we go for the second option, which is to use conventional energy meters, the question is OK, how do I integrate them into the KNX system? Perfect, it turns out that most power meters and network analyzers have the option to communicate via ModBus. We will then look for a ModBus Gateway to KNX, which is indeed easily found.

What the Gateway will do is take a variable from the energy meters via Modbus and send it to the KNX system using the twisted pair, let's assume that both the Modbus network and the KNX network are 100% wired and the Gateway physically has a port for connect to ModBus and another to connect to

KNX. This process of taking a variable from one system and sending it to a different one is called mapping. To successfully do this variable mapping we have to be clear about how Modbus and KNX work, we already know that KNX we have group objects that arc associated with group addresses, so whatever we bring from outside will have to become an object group and associate with a group address.

On the Modbus side we have addresses that refer to the equipment and registers where the information is stored, therefore what we will do is take a Modbus register and make it a group object that we will associate with a group address. So we need to know in which Modbus register is the variable that we need, this information is called register map and it is part of the technical information provided by the meter manufacturer.

Imagine that after viewing the Modbus registers map, we find that the meter with Modbus address "X" stores the overvoltage alarm in register "A" and we want that when this alarm is activated, an emergency luminaire that we have in the address group "Y". So what we will do within the Gateway software, Gateways are usually configured outside the ETS, is to take register "A" from address "X" and forward it as a bit to group address "Y".

Depending on the protocol that is being integrated into KNX, the names will change, perhaps they are no longer registers but universes and they are not addresses but identifiers but the process is the same: You analyze what you want to do, you identify the variable you want to send to KNX , you identify the group address to which you want to send the information, you define the type of data it should have and associate the variables within the Gateway. It works exactly the same in the opposite direction, that is, sending information from KNX to the other protocol or system.

And even better, some Gateways have libraries developed by the manufacturer of the Gateway, saving you a lot of the work but, well, those details will depend on each individual equipment.

Just as these concepts are fundamental, it is also very important to have some technical considerations about the physical installation of a KNX system. We will see them below.

12

Installation

We are going to start with the installation regarding the twisted pair.

The first thing to be very clear is that in KNX the power (110VAC / 230VAC) and the communication (30VDC) are separate, that is, on the one hand you will have the cables that carry current for lighting for example and on the other hand you will have the cable that makes the communication of the teams. It is also important to keep in mind that the most recommended is that the KNX power supply, which will supply the communication bus, be powered by a UPS both for protection of the equipment and for backup in case of failure in the main electrical network.

The communications bus also powers most KNX sensors directly, however some will require additional power as we saw earlier. This is essential to keep in mind because we mainly have two types of KNX cables, 2-wire and 4-wire.In case we only need to communicate equipment without additional power then we will use 2-wire cable while if we need additional power we will use 4-wire cable. Of course, they can be mixed, that is, in one way use 2-wire cable and then in another use 4-wire cable.

Knowing what type of cable we use, we need connectors for that cable, although KNX equipment comes with its KNX connector, but it is recommended to have additional connectors for installation. Remember that although you can use a different cable than the certificate and different connectors than the certificates, this will directly affect the stability of the system and its robustness.

With this clear, we go on to define which electrical boards we are going to use, where to locate them within the installation and how we are going to interconnect the equipment that is outside the boards. For example, we can decide to locate an electrical panel on each floor, interconnect them by twisted pair through the building duct, and use additional piping to connect the equipment found outside the panel on each floor.

With these decisions in mind, at a high level, we begin to have some important considerations. The first is to avoid KNX cable splices outside KNX equipment, that is, each KNX equipment has a connector where we can interconnect 4 KNX cables so we avoid generating points where it is necessary to connect more than 4 cables because it will then be inevitable to make a splice . It is also important to keep in mind that making splices outside KNX equipment can be a problem, for example, imagine a room where we have 4 buttons and it occurs to us to get a cable to the ceiling in the center of the room and from there go down to the 4 buttons. In principle this sounds good and is easier, however it can cause long-term problems for two reasons. The first of them is that we will have to leave an inspection box on the roof because we cannot cover that joint, in case it is released for any reason it must be accessible. And on the other hand we have electromagnetic noise, keep in mind that the KNX cable is twisted and it is also shielded so it is very difficult for electromagnetic noise to enter the communication bus but when we make a splice that protection breaks in the splice and a point of vulnerability is generated. It may sound a bit paranoid but it happens and the truth is that it is better to make the installation as robust as possible.

With this clear and bearing in mind that the KNX cable is identified with colors just like the connectors, the communication bus being red (+) / black (-) and the additional power yellow (+) / white (-). We proceed to verify some important distances.

In case we have two power supplies on the same line, the minimum distance will depend on the particular source that is being used, for this it is important to refer to the specific manufacturer's technical specifications. Later we check that the maximum distance between the source and a non-powered device is 350m, and the maximum distance between the two devices will then be 700m. Finally the maximum length of the twisted pair cable must be 1000m.

Since we have an overview of what our installation we proceed to review some more specific details starting with the equipment that goes outside our electrical panels. For example, buttons, screens and thermostats, it is important to check the type of electrical boxes that the equipment requires, since depending on the country where you are, they may be different from conventional boxes.

Inside the electrical panel some details are important to keep in mind. Starting by marking all the cables and connection terminals, this is essential as it allows easy identification of the equipment and cables for both start-up and future maintenance. It is also important to avoid loops on both the power side and the communications side as any discharge can spread and cause additional damage.

On this point we will delve a little more into the practical part so we will leave it up to here which are the basic theoretical concepts of the installation.

Already at this point we have a high-level and very complete overview of KNX so we will go into detail on each of the subsystems that we can control.

13

Lighting

Lighting was one of the first systems to automate in home automation and automation. Here we find different technologies and different ways to control them. In the most basic lighting systems, only the lighting is switched on or off, in the most sophisticated systems the lighting is regulated and atmospheres are created.

Relays, or equipment with relays embedded in them, are often used to control the lighting on and off easily. For regulation there are different alternatives:

• **Phase regulation**: There are different types of lighting on the market and there are also different types of phase cut regulators, it is important to use the one corresponding to each technology. For incandescent lighting of tungsten filament and halogen lamps, a leading edge dimmer is recommended (figure left). For "Magnetic Low Voltage" (MLV) lighting representing inductive load, it is recommended to use a regulator with leading edge control (figure left). While for "Electronic Low

Voltage" (ELV) lighting, a regulator with trailing edge control is recommended (figure right). This is why in the lighting control industry there are a wide variety of phase regulators, those that are universal is because they adapt the type of control to the type of load they are handling. The regulation by phase has its main application in the domestic sphere.

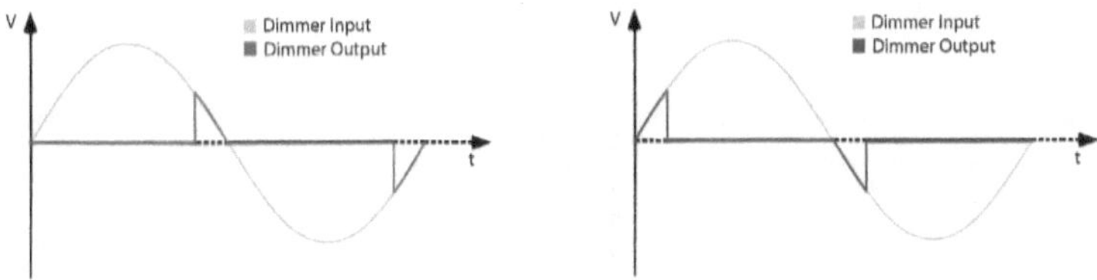

• **0-10V:** This was one of the first electronic solutions for lighting control and regulation, also one of the simplest. For this, a 0-10V ballast is connected to the lighting to be controlled, the ballast has as input a signal that can range between 0-10VDC and this is in charge of regulating the luminaire proportionally to said signal. The typical connection of a 0-10V system is detailed in the figure. The controller illustrated there is a 0-10V single channel controller, that is, it has a single 0-10V output, which means that all the ballasts connected to it are controlled in the same way, therefore, in terms of infrastructure, it is necessary to lay a bus by each independent circuit that you want to handle. At 0-10V it is not possible to differentiate between ballasts, only between channels. In the market there are actuators with 8 channels 0-10V or more, this implies that the same number of groups can be controlled independently. This type of lighting control is widely used in corridors, warehouses or homogeneous spaces where lighting is required to be at the same level. The cost of the ballasts is relatively cheap therefore it is a very efficient solution for this type of space. It is important to note that the 0-10V controllers have a relay for each channel, this is to supply all the ballasts from there and ensure the total shutdown of the luminaires.

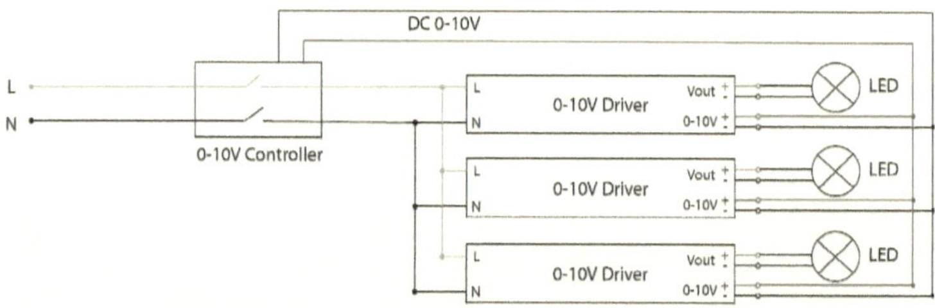

• **DALI:** DALI, "Digital Addressable Lighting Interface", also handles a bus like 0-10V, which travels through different devices with the great advantage of being able to differentiate each device connected to the bus by means of a digital address. This is one of the most complete lighting control protocols and easiest to install. The wiring is very simple since it is a bus that runs through the building and to which all devices are connected, with a maximum of 64 per bus. Additionally, it offers great options such as real-time diagnosis of lighting and its drivers, the possibility of making digital groups, connecting emergency lighting, sensors and keypads. This system is widely used in auditoriums, theaters, stadiums, large spaces with different environments, facades and those places where more specialized lighting control is required. It is also an international standard IEC 62386 and is open so different manufacturers develop devices under this technology. Although DALI ballasts are significantly more expensive than 0-10V, installation and wiring is usually cheaper for large buildings.

The choice of one technology or another for lighting control depends mainly on the particular project and its conditions. For example, for warehouses, warehouses or shopping centers, 0-10V technology is usually used, while for auditoriums or airports, DALI is usually used. In the domestic sphere it depends on the area to be illuminated and the preferences of the inhabitants, but the most common is regular by phase.

Automation of lighting offers great possibilities to satisfy the preferences of the occupants or needs of the building. Among the functionalities that smart homes usually have, the following stand out:

• Time programming: This type of control consists of automating the switching on or regulation of lighting according to the date or time. It is usually used in facades, offices and outdoor areas.

• Presence simulation: This type of control aims to simulate the occupation of the building mainly due to security issues. It is programmed depending on certain parameters on, off or scenarios in the different lights of the building when it is unoccupied to persuade possible threats in a preventive way.

• Occupancy control: Here is a wide range of motion and occupancy sensors, with different technologies and technical specifications. The objective is to illuminate the space only when people are there and automatically turn it off when it is unoccupied, as long as the lighting in the place is low.

• Constant lighting control: The objective of this type of control is to maintain a constant level of illumination in the place. Complement the contribution of natural lighting with automatically regulated lighting of the space. It is usually used in workplaces or places where it is always necessary to maintain the same level of lighting, that is, spaces that have natural lighting and a support is required to automatically complement it.

• Scenarios: Scenarios are predetermined sequences or configurations for a specific purpose. For example, in an auditorium, depending on whether it is a conference, a projection or a play, the lighting must have different settings, which are usually predefined.

• Integration to music systems: Lighting through of home automation or automation can be integrated into entertainment or professional music systems, such as DMX. This is used for shows or for mainly illuminated facades.

• Use of multicolour LED lighting: this type of lighting is generally used in a decorative way and can be integrated without any problem with other systems and types of lighting controls.

14

Blinds

Blinds, awnings and some mobile façade elements are a fundamental part of home and building automation. Among the blinds we find different types such as roller blinds and Venetian blinds, however the motor they use is usually the same and only changes the way of controlling it. Regarding awnings, we have exterior awnings and interior divisions to section spaces. And as for the mobile elements of facades we have gates, garages, windows or ventilation gills. Although most of these elements are apparently disparate, the motors used to control them are very similar and the main ways to control these motors are:

• Infra red: Using infrared signal remote controls, usually with a 940nm wavelength. This type of control occurs in some gates and shutters. Its main drawback is that it requires line of sight to be able to be activated.

• Radio frequency: This type of control is very common in devices aimed at the end customer. They usually use the 433mHz frequency. The manufacturers of blinds have developed signal concentrators to make their management easier in medium-sized installations. The main problem is the interference they usually present with each other when the number of devices increases. Mainly seen on shutters and gate or garage motors.

• Communications Bus: This type of control is less common and is usually found in very specific products. For example, high profile and precision gills for facades or blinds.

• Conventional 4-wire motor: This is the most common and usually the most robust control. It consists of 4 wires corresponding to ground, neutral, rise and fall. By feeding the power line, the upload cable generates this action and in the same way it is lowered. It is important in this type of control to make sure by means of interlocks that under no circumstances will the up and down cable be energized at the same time, because this causes damage to the motor. It is also cheaper to order 4-wire motors from manufacturers, however, the installation of the necessary infrastructure for their control is significantly more expensive than a wireless solution like the previous ones.

These elements are used to protect the interior of the building from light, to grant privacy, to preserve the temperature of the building in a natural way, to access it. The control can be done by monitoring the sun by twilight, by time programming, by

scenarios and according to information provided by stations weather such as rain or wind speed.

15

HVAC

Around 40% of the electrical consumption of a building corresponds to heating, ventilation and air conditioning (HVAC). This is why it has been a fundamental system in home automation and imotics. There are a wide variety of HVAC equipment, among the most representative are:

• Fan coil: It is a device consisting of a heat exchanger for heating and / or cooling, and a fan. Usually connected to a duct system where hot or cold air circulates as the case may be. It normally has two solenoid valves, one for heating and the other for cooling, and three fan speeds. There are a wide variety of actuators in home automation or immotic to control fan coils, these actuate the solenoid valves and fan speed to control the temperature of the room. They are mainly used in commercial establishments or institutional buildings.

• Split and mini split: These are heating and cooling systems that allow the temperature of individual rooms to be controlled. They have two main components: an external condenser and an internal evaporator connected by a conduit that houses the power and communication cables, the copper pipe and a condensation drain line. It is a system widely used in buildings where its initial design did not include HVAC systems. The control of these is usually by infrared.

• Radiant floor: The radiant floor consists of a system of pipes embedded in the floor, through which hot or cold water flows as the case may be. In cold cities it is widely used since heat transfer is done from the structure and therefore the thermal inertia is high. However, this is also usually a disadvantage since it can take hours to perform a significant temperature change, unlike other systems that require minutes. This system is controlled from solenoid valve concretrators that control the flow of water through the different areas.

• Chimneys: The current chimneys are usually gas, organic fuels or plasma. These are usually controlled by radio frequency and in some cases by infrared signals or even dry contact.

• Fans: Fans are very common to reduce the temperature of a certain space. The control of these is usually done by regulating the power signal that is delivered to it. Those that already have remote control systems are usually managed by infrared or radio frequency signal.

In the different open and closed protocols there are specific actuators for each type of HVAC system, as well as thermostats and management systems. Among the most relevant controllers are:

• 2-point controllers: In these controllers we define two setpoints that we will call "a" and "b". Now if the current temperature value is below "a" then the actuator is activated and if the current value is above "b" then it will turn off. This is the simplest type of controller and its advantage is that it is fast-acting but with only two levels the control precision is low. For example a refrigerator or a hot water heater.

• 3-point controller: It is very similar to the 2-point controller but it can work in two modes, heating or cooling. Here if the current value is below the lower limit then it turns on the heating and if it is above the upper limit then it turns on the cooling, while in the middle it turns off both the heating and the cooling. These types of controllers are simple and their reaction is still fast, the control precision is better than the 2-point controller but their consumption is higher. Also, if not configured correctly, considerable energy waste can be generated. For example a simple room controller with heating and cooling mode.

• Proportional Controller: As its name says it acts proportionally to the deviation of the desired point, that is, the larger the deviation the greater the performance of the controller. This type of controller is intermediate reaction but it has some disadvantages such as its high sensitivity to noise and its error in remaining in stable state. For example, a water level controller in a bathroom, although the truth is that this type of controller is rarely used only due to its error in the stable state.

• Integral Controller: This controller has an integral action, which means that the longer the deviation lasts, the greater the performance of the controller, because the longer the error, the greater the area under the curve, therefore the integral. Therefore this type of controller is used to eliminate the error in stable state, the same error that the proportional controller leaves, so it is usually used in conjunction with a proportional controller. The problem with this controller is that fast deviations escape you.

• Derivative Controller: This type of controller has a derivative action which means that it reacts to variations, the more sudden the variation, the greater the action of the controller. It is usually used in positioning and stability control.

• Composite PI / PD / PID controller: This type of controller is the most widely used in real life, since proportional, derivative or integral effects are combined depending on the system to be controlled. KNX temperature controllers typically use a combination of proportional and integral effect, the proportional for fast action and stability supplemented with an integral action to eliminate error in steady state.

16

Safety

One of the most important fields of action for home automation is that it helps guarantee the safety of material assets and even more important the safety of the people who live in said space. Thus returning to the architectural structure an ally in favor of the care of the most precious assets.

Due to the development of the different protocols, it is possible to integrate different security and alarm functions that already exist in the market.

Systems focused on security in the physical structure of buildings and homes can basically be categorized into four areas:
   • Intrusion alarms for intruder detection. (Movement, presence, pressure, etc.)
   • Technical alarms. (Fire, smoke, flood / water, gas, power failures, telephone network outages, etc.) These types of alarms by themselves only detect the failure, however, together with an

automation system, not only detect, but the system knows how to deal with these failures and / or threats, performing in many of these cases actions that are subject to human participation, however, our home automation network can perfectly execute them to take care of everyone's well-being. more effective.

• Personal alarms. (Panic and assistance systems) Silent alarms that allow users to send alerts to those who can react more favorably to the situation. (Central stations for alarms, police, private security, etc.)

• Video surveillance. (CCTV and IP). It includes some systems focused on supervision and action on security against intruders, access controls (biometric, electronic, electromechanical, among others.) And emotional video surveillance.

Regarding the human security approach, tasks such as:
• Lighting common areas for passages and accesses that may be risky stand out. (Stairs, country terrain, etc.)

• Deactivation of power plugs and sockets.
• Manipulate systems remotely.
• Urgent emergency notices discreetly and safely.
• Detectors for leaks or damage to closures and valves that affect the life and integrity of members of the home and tangible assets.
• Tele service and tele assistance to people with special needs.

In all areas you can include various sensor's and installations that complement the security services and the actions associated with them.

In general terms, the security systems associated with the home automation area and that include services for it, if they are well managed by automation for the sake of creating true intelligent buildings, become the most reliable ally to put the security of what more we want.

Domotics systems are not subject to using a single protocol for their communication, however it is more convenient to include endorsed equipment for such functions and certificates in their respective tasks.

17

Audio/Video

In home and building automation, the entertainment system mainly consists of:
• Audio: Distributed audio systems allow different music to be heard at the same time in the different spaces of the building. Likewise, the same audio can be reproduced in the entire building or made peripheral. To achieve this, audio matrices with several inputs and several outputs are used, which can be addressed by programming or by remote controls.

• Video: Video systems can also be distributed using video matrices with different inputs and multiple outputs. The control of these systems is usually done through infrared controls, through the hdmi cable or by programming.

Let's start by clarifying some key audio concepts. We have all listened to surround sound systems or we have even bought a sound system or a 5.1 home theater but few of us understand what that means, so let's look at the different multi-channel audio configurations we have:

• Audio 2.1 : System with two channels, stereo sound and an additional channel for low frequencies.

• Audio 3.0-3.1: Systems with three front channels (left, right and center) or two front channels (left and right) and one center rear channel. The 3.1 has an additional channel for low frequencies.

• Audio 4.0 - 4.1: System with three front speakers and one rear. Or two front and two rear speakers, quadraphonic sound. The 4.1 has an additional channel for low frequencies.

• Audio 5.1: System with three front speakers, two rear speakers and an additional channel for low frequencies.

• Audio 6.1: System with three front speakers, two side speakers, one rear and an additional channel for low frequencies.

• Audio 7.1 - 7.2: System with three front, two lateral and two rear speakers. While 7.1 has a single additional low-frequency channel, 7.2 has two.

• Audio 8.1: System with three front speakers, two side speakers, three rear speakers and an additional channel for low frequencies.

• Audio 9.1: System with three front, two side, three rear speakers, one ceiling and an additional channel for low frequencies.

All of these systems are made up of speakers, so it is important to know how to identify and select good speakers. Here the first thing to keep in mind is so that application you want the speakers, that is, if it is for a home where they only require background music or if it is for example for a theater in a house where the client is audiophile. The approaches are totally different and depending on the level that is required in the audio system and of course the budget, you can limit the selection of the speakers.

The first factor that determines the sound quality of the speaker is the design and construction of the transducers, that is, the part that moves and converts the electrical signal into audible vibrations. Although two speakers are very similar at first glance, the materials and the typology of the transducers generates a significant change in the sound and therefore in the price. In this, the most important thing to take into account is the frequency they are going to reproduce and the power, since while low frequencies require larger transducers and cone type, high frequencies will require smaller sizes and dome type.

If a type of transducer should be used depending on the frequency, then the immediate question is how this is applied in real life, because here we enter the second factor, which is the frequency filters. These are electronic components that literally separate the input signal at different frequencies. Most commonly, one or two cut-off points are handled, generating two or three blocks of frequencies known as the "loudspeaker pathways". These filters can be active or passive, the passive ones are those commonly found in general application loudspeakers and the active ones in high-end ones. They are called passive because they do not require power whereas the active ones do require it and are usually involved in the amplification process.

And the third main aspect is the speaker box. Here, just as musical instruments, the material is fundamental, a chipboard is not the same as a speaker with solid wood, the sound quality and the price are very different. Shape and design are also important, including the speaker's internal sound insulation.

Regarding the video, it is important to be clear that it is a smart TV. Basically it is a television with an operating system, yes, an operating system similar to that of your computer or your cell phone. This has great implications for you as an integrator. The first is that it opens the doors to control the television by means other than the typical infrared control, this is because having an operating system, the television makes it viable to develop applications and drivers that allow the television to be controlled through the local network, even you can have the feedback that does not give an infrared remote control. It is a matter of time for television manufacturers and KNX equipment manufacturers to bring to market products that control smart TVs through the local building network.

18

Measurement and energy efficiency

A well designed, programmed and executed home automation system leads to savings in electrical energy, associated with the basic installation systems: lighting, air conditioning and saving resources such as water; These are common factors in buildings on which automation can have a great impact to control, monitor and regulate them in the most optimal and efficient way.

Most of the savings in a building can be achieved by using sensors and proper programming that limits excessive use by the necessary ones.

• Measurement of water consumption: Using different sensors, manage to control the water consumption associated with irrigation systems. , efficient taps, closings due to breakdowns and measurement of level systems to prevent damage from floods.

• Measurement of electrical consumption: It is possible to know accurately the consumptions that are necessary from those that are not necessary to achieve adjustments in the controls of the building's systems.

• BMS: Use of specialized software for the analysis, recording and control of measured data, with the aim of establishing continuous improvements.

One of the most important factors when automating a residential or corporate building is to rely on measurement studies, since these can greatly optimize the integration results; Only what has been measured can really be improved.

Regarding energy, there are a couple of important concepts to be clear about. Starting with the different types of electrical power. First of all we have the active power, also known as average or real power, this is generated by the resistive devices and it is the power that is really used as useful power, it is measured in watts (W). Secondly we have the reactive power, this is the one used by the coils and capacitors to create electromagnetic fields, as such it does not become effective work and is fluctuating in the network, it is measured in reactive voltages (VAr). Finally we have the apparent power that is the total power consumed by the load and is obtained from the vector sum of the active and reactive power, it is measured in volt amperes (VA).

The next concept to keep in mind is harmonics. Harmonics are higher frequency waves that are added to the fundamental wave, which are always multiples of the fundamental frequency. That is to say, they are, after all, of higher frequency electrical noise found on the fundamental electricity wave. Harmonics are often thought to be introduced by factors external to the network, but the truth is that they are created within the network by non-linear loads.

The third fundamental concept in the measurement of electrical energy is the precision of the meters. Here they are classified by the limit of permissible percentage error in the measurement for all current values between 0.1 or 0.05 times the nominal current and the maximum current. Class 0.2, 0.2s, 0.5, 0.5s, 1 and 2 are conventionally operated, the class 0.2s meter being the most accurate.

Well now referring to the measurement of consumption of water or other fluids there are also a couple of concepts to be clear. Starting with the flow, which is the volume of the fluid that passes through a defined area in a specific time, this is very important because each meter has a defined flow range and depending on the expected flow, one meter or another will be appropriate.

The second aspect to keep in mind is the type of fluid. Because due to the nature of construction and measurement of the different types of meters, the type of fluid may or may not affect the measurement itself. For example, in magnetic meters, conductive fluids can be measured, while "orifice plate" or "vortex sheeders" meters are not suitable. On the other hand, in most turbine gauges, steam cannot be measured, while in most vortex and differential pressure gauges, liquid, gas and steam can be measured.

And finally the third concept to keep in mind is the flow profile that depends on the nature of the fluid. This can be classified into laminar, turbulent and transitional. The laminate makes the fluid flow in the center of the pipe faster than in the walls, practically in layers and is due to the viscous force. Turbulent has the characteristic that the inertia of the fluid is greater than its viscous forces, thus making the flow more uniform, however the layer of the pipe wall continues to travel a little slower. Finally the transitional, is between the laminar and the turbulent, oscillating between them and making the fluid more unpredictable.

19

Integration

As we have seen, there are different systems within an intelligent building and therefore their integration is essential. Likewise, the interaction of these systems is essential for the proper functioning of an intelligent building.

The integration of building systems refers to the intercommunication between them and centralized control in one place. This can be done in different ways depending on the particular project.

In home automation, central controllers are usually used that integrate the different building systems into one application, making it easy for the user to operate and allowing the systems to communicate with each other. The objective of integration, apart from centralized control, is to increase the functionality of the building through shared scenarios or functions. For example, a stage called "cinema" could turn off the lights, lower the blinds, unfold the television from the ceiling, direct the sound of the television to the audio system and open the list of stored movies. An alarm scenario in a smart building could, apart from activating the alarm, turn on the lights of the sector where a possible threat is detected, raise the blinds, take a photo, send it by mail and allow the owner to make a security message to through the audio system the entire building.

When the buildings are larger and the systems to be integrated are more complex, a "Building Management System" or BMS is used. These are systems that require a specific computer or electronic card for installation and handle different protocols to control the other building systems through them. Most BMS work under the BACnet protocol and some include other protocols like KNX, Modbus, JSon and SNMP.

Once the integration system is installed, remote access to the building is almost imperative. For this the most used options are fixed IP, dynamic DNS, VPN or own servers of the integration system.

Current integration systems beyond making smart houses make automated houses. Where depending on parameters such as the schedule, presence of rain, movement, wind speed, temperature and security intrusions certain automated actions are performed. Most of the control has to be done by the person and the level of abstraction of the building is very low. So many BMS provide the option of using libraries or creating scripts to increase its functionality. Now we are going to practice!!

III

LET'S GO TO REAL LIFE

The third part of this book is focused on landing the theory previously seen in practice, as well as containing important considerations for moving from theory to real life. In particular this part of the book I recommend to be read every time you are doing a new project because each project is different and you will find attuned to different tips every time.

"It is not enough to achieve wisdom, it is necessary to know how to use it." Ciceron

20

Why should you use KNX?

At this point we already have the theoretical foundations to get going, which is great because we can speak the same language. And although we have been as specific as possible it may feel somewhat complicated, after all we could automate a building with systems that do not require thinking about a topology, group addresses, group objects, couplers or gateways. Well yes, but no. The first thing we have to think about is the size of the project and its technical requirements, if it is a small project such as an apartment-studio because there is no need to use KNX, with a relatively low budget and without the need for topology or wiring of any kind you can use DIY type IoT home automation equipment ("Do it yourself"). The market has several options in this market range. In the future, when you get bored with these devices at home or want to change them with a comfortable budget, you can change them. That is, we can apply a fair solution that meets specific requirements without the need for KNX. In this case, use KNX, put together a topology, perhaps include different means of communication, and use a mini

BMS because it would be like killing a mosquito with a bazooka or a fragmentation grenade. Even implementing these DIY systems could be done by a customer on their own, without needing you or me.

Now if it is a larger project, the matter begins to get interesting. The first aspect to evaluate is how many systems I want to integrate or automate in the building, both now and in the future. If I only want to include a system already in the future, there is no possibility that more systems will be integrated, because you can surely choose a proprietary option or a solution from a particular manufacturer, although well let's be realistic in the world in which we live, chances are that Sooner or later all buildings go to be largely automated. Therefore, the healthiest both for you as an integrator and for the end customer is to opt for an open and multi-manufacturer system. Here KNX enters the scene, since it is an open protocol with a trajectory in the market that provides guarantee and future projection in interoperability with more than 350 manufacturers and a presence in practically the whole world. Associated with this, you have a wide variety of Gateways to integrate both proprietary and open protocols, even if you use KNX as the central axis of the project, you can make two proprietary systems that in principle have no way to interact do so through KNX.

The second aspect to take into account is the regulations that will support the system used. KNX complies with different international regulations and is considered a world standard, which gives you enough support and security, to be precise it complies with: ISO / IEC 14543-3, EN 50090,

EN 13321-, EN1332-2, GB / T 20965 and ANSI / ASHRAE 135. Additionally, it is important to keep in mind that all KNX products independent of brands or manufacturers go through a certification process with a testing laboratory, therefore you have the guarantee that any KNX certified equipment Regardless of the manufacturer, it meets technical specifications for both operation and interoperability.

The third very important aspect to keep in mind is the use of a single software for programming the KNX equipment independent of the manufacturer in question. That is, if you have 5 different manufacturers in your project, you only need one software, the ETS, with a single license for everything. This is a great advantage because in other systems you have software for each brand, which makes programming and commissioning absolutely tedious and expensive both in time and in licenses.

The fourth relevant aspect is the particular installation and the transmission media that you will need to carry out the integration of the systems. With KNX as we saw some chapters ago you have different means of transmission that can be combined with each other and manage to meet both customer expectations and technical requirements, optimizing civil intervention in the building of course.

Finally, the support that the client and the project will have in the future is relevant. With KNX, as well as there are more than 7000 products from different manufacturers, there are also several companies around the world with certified KNX personnel, this implies that if tomorrow you you win the big prize of the lottery and you decide to go to live in Fiji then the end customer will be able to find service support with another company or integrator. It is important to think about the end customer and that the system you implement lasts over time with the best guarantee in both support and equipment.

At this point you already have the necessary tools to discern if KNX is the best option for a project and if so, then surely you will be able to convince the client of why it is the best option. Since the decision is made, let's get to work!!

21

How to get projects and clients

With the clear fundamental theory and the reasons why to use KNX, as well as when to use it, the immediate question is how to get projects. Because without projects all this will remain in theory, if you are an integrator you need projects to put this into practice and to live in fact. But in the real world getting projects requires more than certification or even being good at KNX.

Here we will not see those projects that literally give away for simple convenience or even those that fall by luck, here we will address strategies and tips to win the projects, to make merit for them and to increase the chances of obtaining them.

The first thing is that you have to be good and master the subject, this does not imply that you master all the details of the implementation or the minor ones of the technical detail. This implies that you master the subject at least at a high level and have clarity of the route to design, specify, implement and do post service sale of a project. It is important that you master the theory presented here to have tools that will allow you to enter the market.

Now to enter the market, assuming you have just left your KNX courses, you have a couple of options for it.

The first we will call "Prospecting". Here the objective is to find the clients yourself, for which the initial step is to define the client. For this I recommend making a business model canvas with all the elements that are required, it may seem boring or tedious, it might even seem unnecessary but it is really very important since it will allow you to land things. For this you basically need to ask four questions:

• Who ?: Here the objective is to define who your customers will be. It is essential to define a market segment to target, so then go to another but it is important to make it so specific that you can create your ideal customer. You must get to the point of having a single customer profile, with a certain tolerance, of course, which will be your ideal customer, understanding as an ideal customer the one who buys you at the price you ask for and pays you on the terms you choose. Once you have defined your market segment it is important to define two complementary aspects. The first is how you will manage relationships with your customers and the second will be your distribution channels to reach them.

• What ?: Here it is important to define your value proposition and that includes that it differentiates you from the rest of KNX partners, even from engineering or automation companies who, even if they are not certified in KNX, can win you the

projects and then certify one of their engineers. So it is

important to do a little reflection and put yourself in the client's place, because he should choose you and he will win with that. If you have a couple of reasons why the customer wins by choosing you and not your competition, or what you can do that your competition cannot do, then you have this part ready.

• How ?: Already with whom and who we go to as. Here are three main things to define. The first will be the key resources you need to do the projects, for example the ETS license or a draftsman to do the plans. The second will be the key activities you need, for example electrical design and ETS programming. And the third will be the key associations that you need, for example, your equipment distributors or if you need to make a consortium with a company that provides you with economic strength, you need a distributor or several that support you with the supply of the equipment.

• How much ?: Finally we turn to the quant, to the business numbers. Here two aspects are the main ones. First your operation and customer acquisition costs, fixed and variable costs. Second, your source of bone income has to define how much you are going to charge and how you are going to do it.

These elements are reflected in a table that occupies a page, here we will have at a glance all the elements of the business. This step is essential whether you are or want to be an entrepreneur or if you are already in an established company, this canvas is a compass because without this you will most likely lose drifting.

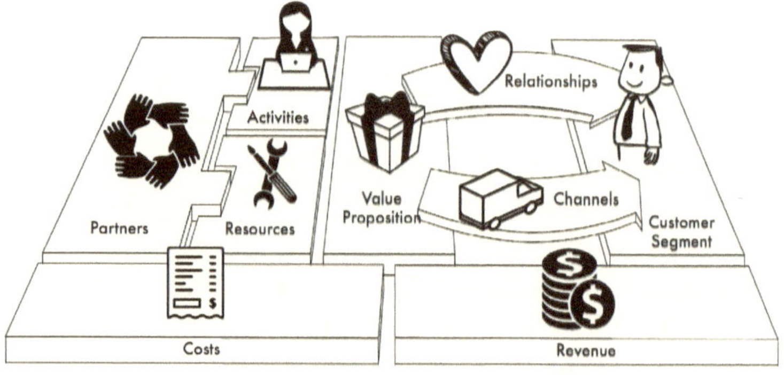

Once you have captured the business model canvas, you can see if your model is viable and if it is, we will move on to the next stage, which is to zoom in on the value proposal, that is, we will structure in more detail our value proposition. For this we have to do two main analyzes, the first is the customer profile and the second is the value map.

We will start by doing the analysis of the customer profile. For this we will have to focus on the client and understand three fundamental aspects:

• Jobs to be done: We will define what jobs or tasks our client has to perform in their daily life or work. Here we will have functional jobs, that is, those things that our client does to fulfill a specific task or solve a particular problem such as making quotes for automation systems. We will have second social jobs which are those that our client does to make a positive impression on people such as being perceived as young and innovative. Thirdly, there are the personal jobs that are those that our client carries out to feel good, such as installing a security system to feel safe. Fourth, we have the support jobs that are those that our client carries out in the process of consuming value, for example buying products, participating in forums or reselling products that they do not use. Once we have the works of our clients identified, we will have to order them according to their importance for the client, eye is the importance for the client, not for us.

• Frustrations or pains: They are those things that prevent our client from carrying out their work or tasks to achieve their desired results. Here we will have three types of frustrations or pains. Firstly, unwanted things, whether they are results, problems or product characteristics, for example making quotes is long and tedious. Secondly, we will have the obstacles that are those things that prevent our clients from completing the tasks that they have to do, for example to make a quote for a system, the engineer has to do a design and is out of the country. Thirdly, we have the risks, those things that could potentially lead our clients to results or unwanted things, such as if the quoted system is not correct, it may incur over-stitching or loss.

• Gains or joys: These are the results or benefits that our client wants. Here we will have in the first place the required profits, that is, those that are compulsory for example an automation system has to work correctly all the time. Secondly, there are the expected profits, that is, those that are not mandatory but that the customer expects, such as an automation system, that is expected to be intuitive to handle. Thirdly, there are the desired profits, that is, those that would make our client happy and that they would like to have, for example, that the automation system is fully integrated with our virtual assistant. And finally there are the unexpected profits, that is to say, those that our client does not even expect but that would make him very happy, such as for example that the problems that arise in the automation system are detected and corrected automatically.

Once we have identified the jobs, pains and earnings of our clients, we will have to order them according to their importance or relevance to the client. With this we complete the client's profile and capture it in a drawing.

Now we move on to map the value of our product or service. For this we will have to analyze three things again:

• Products and services: This is simply a list of what you offer to your customers. You can offer physical or tangible products, such as a touch screen. You can offer intangible products as after-sales support. You can offer digital products or services such as online programs or recommendations. You can also offer financial products such as financing or payment facilities in projects.

• Analgesics: Here we describe in an explicit and very specific way how we are going to alleviate the pain or frustration of our clients. In this part it is very important to make it as explicit as possible, for example we are going to alleviate the frustration that the quotes are tedious by offering our client an online platform where you can select the number of rooms and the subsystems to automate, choose the products and how As a result, you will obtain a detailed list of the necessary components as well as design, installation and programming services with their respective prices.

• Profit creators: Here we will have to describe, again in a very specific way, how our products or services generate profits for customers. For example, designing a display system with a maximum of 5 buttons per screen and a navigation system such that with 3 clicks you can go anywhere, we will make our client have an intuitive and easy-to-use system. Here I reiterate it is very important to be as specific as possible. Once we have these three aspects analyzed, we will proceed to make a classification in each section according to the importance and relevance for our client. We will capture this again in a drawing.

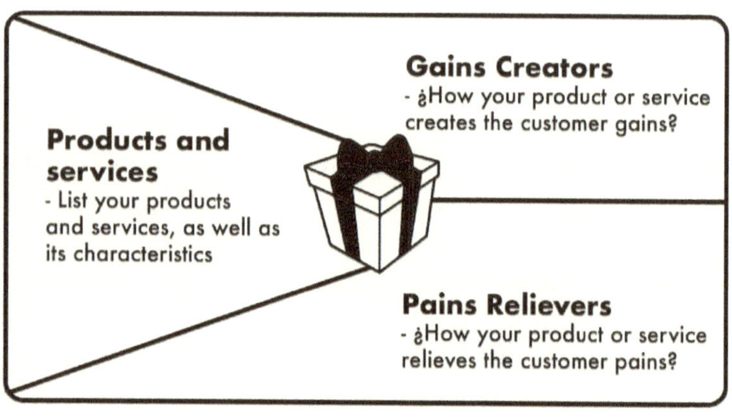

The next step is to unite our client profile with our value map, putting one in front of the other and relating the elements of each of them. We will relate the creators of profit to the profits or joys of our clients and we will relate the painkillers to the frustrations or pains of our clients. It is important to keep in mind that no product will ever solve everything for everyone, that is, the fundamental thing is to focus on the most important pains and gains for our client.

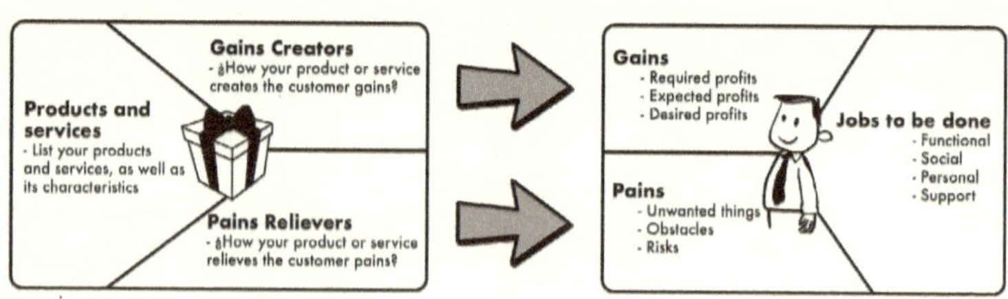

At this point you will have the necessary information to go

looking for your clients. You will also have the information you need to be able to define what type of advertising you will do and through which media. This will allow you to make the most efficient use of the economic resources you have for the acquisition of clients. The truth is that this option is the most difficult and going hunting can only take time, patience and a lot of work.

The second option to get clients may be the most efficient and requires a paradigm shift at least temporarily. Yes, it is true that KNX is a multi-manufacturer and the ideal is to use the best of each manufacturer to achieve the best combination of equipment that provides the greatest benefit to your client, but yes, life is far from ideal. What you can do is identify a manufacturer that is very influential with which you can make an agreement to work with, that will help you start and you will use their products for your projects. The truth is that for the vast majority of KNX projects, any large manufacturer will almost be able to offer you a complete solution with their products, as well as, which is beneficial for the manufacturer because it will increase its penetration in the market and its products even the least sold and it will be beneficial for you because many of the projects really reach the manufacturers first and it is they who direct it as it is most convenient. If you are starting you will have to make a reputation with that manufacturer and have a very good level so that the first project that you trust is perfect, also keep in mind that the dream of a manufacturer is to have distributors or integrators so good that they can really solve almost all the problems autonomously. The worst nightmare for a manufacturer is having integrators or distributors that give them more problems than necessary and the projects they execute work in an not optimal way since it is the image of the brand that is compromised. If you opt for this option of looking for a manufacturer to leverage you, then with that in mind you are going to do the business model canvas and the value proposition with this new approach again.

If you already have projects, I recommend that you carry out the business model canvas in order to focus your economic and human resources optimally, I assure you that you will be able to increase your projects and therefore your profits.

The other side of the coin is that you are a manufacturer or work in a company that makes KNX equipment and distributes it through channels. In your case the important thing is to get really good integrators that make your life easier instead of giving you a lot of unnecessary problems.

Here are several things to keep in mind. The first thing is that a good integrator can be done, that is, you can perfectly train people since you have both the knowledge and the staff to do it. But there are things that can spoil that ideal. The first of them is the integrators company staff turnover, if the staff turnover is high you can think that those who left are bad and maybe this is the case, it is an option if they are one or two, but if the turnover of staff is high so it is not a problem of people but of the company as an organization. This will be a big problem since you can train every engineer that gets there, but they will go away and you will have to start over and over again. The managerial and administrative part of the integrating or distributing company can ruin your training and qualification efforts. So the first thing is to select a company that has good staff permanence and where the workers are comfortable. If it meets that, we can continue, if it does not, then that integrator will be a headache for sure.

If you are going to choose to train the integrator it is important to keep in mind the learning curve and during this curve if the integrator does not have additional income, he may die in the process. In any case, the best option is to deliver small projects and increase the difficulty, including the amount, as they are completed. Delivering a large project to an integrator who is not sufficiently qualified is a bad idea and can burn you, which will naturally have an impact on your profits. Thus, the more qualified the integrator, the better, that includes all the aspects of the business model canvas already described, for example, an integrator without a well defined supply chain or structured project management processes will hardly end a project in the expected time with the desired quality. So the next step is for the integrator to be very clear about their business model and value proposition, to align it with your goals and policies as a manufacturing company and thus generate stable mutualism.

22

Good design practices

Let's start with the first, identify the client and make a profile of it. Surely you did not expect that, maybe you thought that the first thing was technical specifications or the plans of the building ... well luckily you are reading this book because this is not taught in the class of electrical networks or automation.

Identifying who your client is and making a profile of it is essential, this segmentation will give you a clear horizon in the design. Remember that the project is not for you, it is for your client and quite possibly has absolutely different preferences or needs than yours. Imagine that you are making a summer house for a couple of young entrepreneurs with a baby who is in the process of starting to crawl, surely it is different than if they are a couple of pensioners without small children. For the first couple maybe 4-key buttons with different functions on each key or even two functions per key will be a good idea and they would like to handle everything on the phone without remote controls, but this maybe not is the best idea for the second couple who prefers to have only 1 function per key with a large easy-to-see symbol and remote controls for operating the system because the cell phone is not something they like to carry with them all the time. The system is fully functional and it may be optimal for you as an engineer, but remember the system is not for you, it is for your clients and understanding who they are, as well as their preferences, is essential to guide the design.

Once you are clear about who your client is and you have a well-developed profile, you have analyzed your preferences and defined your expectations, we proceed now to think about what systems we need to integrate. We go from the most general to the most specific, at this point we are only interested in analyzing if they are lighting and HVAC, only access control, perhaps multimedia and security, etc. Already having clear that systems we are going to integrate or automate, we are going to contrast the customer profile with the building to make a projection of which systems could be integrated in the future. The systems that we identify that can be integrated in the future will be included in the same list of those that will be integrated immediately. Yes, on the same list.

With the list of the systems that we are going to integrate, we proceed to go into the detail of the building, analyze its structure and define if within it there are particular needs to start designing the topology at a high level. That is, we are going to imagine that it is a 6-story multi family house where on each floor a son lives with his family and on the last two floors parents live, here we have to identify and profile each of the families individually. We can think for example in generating a line for each apartment and joining them with a main line by means of KNX / IP couplers. We could start counting now and use line segments to join perhaps two floors and save on making an additional line, but this is called good design practice not to save costs. The idea is to make the project as organized as possible and also using a different line for each apartment allows us to better manage telegram traffic, remember that the couplers work as filters.

Now we will go to the apartments in more detail. We will analyze the different areas within the apartment and what systems from our list are present, or could be present in it. Then we will analyze system by system keeping in mind that functions are in daily or more common use, and that other functions are more specific taking into account the customer profile. When finished, the result will be a table where for each room we will have the systems that from our list, with the functions of the same in that particular room, differentiated from the most daily ones of the specific ones, we will call this automation matrix. Based on this table, we will design scenarios that integrate the different systems at a high level, that is, here we are not talking about objects and group addresses, we are only doing the conceptual design of the project. At this point we will have a fairly complete conceptual design of the project.

With this, we can analyze which equipment will be more functionally appropriate for us, because we already know that both aesthetics and technical operation are very important and either one can be a design criterion depending on the customer profile made at beginning. So let's move on to the specifications of the equipment.

23

Successful specification

For this point we have all the theoretical foundations and the conceptual design structured enough to face a correct specification, an essential part of an engineering project of any kind. We are doing home and building automation professionally so we will proceed as such.

Again we are going to go from macro to micro. We begin by analyzing the topology and system integration requirements, as well as the required visualization or management system. With this clear we begin to go down to the subareas, for example the apartments, and analyze in them the macro requirements such as number of users, operating systems of the clients' devices, characteristics of the local network of the building, existence of private networks and existing subsystems. Regarding the existing subsystems, it is important to identify if it is feasible to integrate them or if they are really obsolete and need to be replaced. This is important since it will indicate the amount of traffic that the system will have, it will indicate what protocols or systems should be integrated, so we can begin to specify the gateways, it will indicate the type of visualization or management software that we are going to need. It should be noted that this is an iterative process, although we know that protocols must be integrated, we still need to define the number of variables in each protocol and the same for the management system, which we will gradually reveal.

We continue to define the means of communication that we are going to use in the project, this depends on the possibility and the willingness to carry out wiring or civil works, if applicable. The objective is to make use of the twisted pair as much as possible, we have already seen its advantages and the objective is to take advantage of them. It is true that for our part we want to mainly use twisted pair in the projects but on the client side or some architects prefer to use wireless communication means. Therefore it is essential that it is clear that the best option in functionality and support is to use the twisted pair in the lower parts of the topology, since if the associated cost is not possible or absolutely laughable, we will complement the system with wireless means of communication such as radio frequency or IP.

For the backbone or the upper parts of the topology, it is a good idea to use IP since it allows us to better manage traffic, even providing important flexibility to the system.

We continue in the sub-areas analyzing and designing the integration of each of the subsystems independently. Here we begin direct interaction with each of the specialists, keep in mind that each subsystem is a world and unless you thoroughly master each of them, it is best to rely on expert people. And so why did we see the fundamental theory of each subsystem? We saw it because you have to understand each subsystem and understand it well to be able to successfully integrate it and even give high-level guidelines to the particular subsystem specialist. For example, your do not know how to make lighting designs and calculate the luminosity of each space taking into account the color reproduction and the uniform distribution of light, but if you know that there are different types of dimming and that for example for a particular project DALI is the best Option for the flexibility that it provides in installation and control, since the specialist is responsible for making its design with the appropriate luminaires and according to your instructions, use DALI controllers for the luminaires. The same happens with each subsystem, you as an integrator provide the general guidelines resulting from high-level design and you work as a team with the specialist in the implementation of each subsystem with its subsequent integration.

Although it sounds simple, there are some general guidelines that make the job more efficient and result in an accurate specification of the equipment. The first thing is to make sure that each one of the subsystems can be controlled directly from KNX, it has the option of being controlled through a Gateway or a standard communication protocol. Ideally, for example, lighting can be controlled directly with KNX dimmers or actuators, however, if this is not possible, it is important to verify the existence of gateways that allow it to be integrated into KNX or, in the worst case, to use a communication protocol. to control it.

Bearing in mind that the subsystem can be integrated into KNX and defining what type of equipment will be used for this, it is important to define the number of variables that will be handled, as this will make a team suitable or not for this integration. Keeping in mind that a reserve range must always be left, we remember that we had the projection that additional requirements could be generated, since the equipment that we specify must be able to expand its functionality or integrate with new future equipment. That is, if we are going to integrate 100 variables through a Gateway, but according to our projection there could be 300 at some point, since we are going to use the Gateway that is capable of handling 300 variables.

It is very important to keep in mind the nature of the equipment to be controlled and the capabilities of the controlling equipment, for example if we are going to dimmer an LED strip, make sure that the dimmer is adequate both in voltage and power, even in the form of visible dimerization. We will look at some of these details in the following sections.

Let's go for it !!

Lighting:

The most common and practically obligatory system in home and building automation is lighting, so let's start with the big shot. To achieve a correct specification in lighting there are several aspects to keep in mind.

Firstly, the distribution of the circuits of lighting is important, since the number of channels that we will need in our lighting actuators depends on this. Here it is important to find a balance in such a way that the system has the flexibility to generate different scenarios and environments with the lighting but that does not become too complex to use because the effect can be opposite, that is, by leaving too many lighting circuits in In one area the client may feel overwhelmed and end up using just one of two circuits.

Later we analyze the type of luminaire, since depending on the type of luminaire the appropriate dimmer will be one and the other, here I refer to the technology of the luminaire for example incandescent, fluorescent, halogen or LED. Imagining that they all operate at the same voltage and with direct dimming as the technology is different, which implies that the type of load is different and therefore it is essential that the dimmer explicitly say that it supports that particular type of load.

Finally we enter the type of dimerization, operating voltage and current. And note that if we are talking about regulation, the frequency matters, that is, if the dimmer we choose is for 50Hz and our installation is 60Hz because we will surely have problems. Here it is very important to make sure that we are going to keep the phases of the electrical installation balanced on our electrical panels. Although we can connect all the lighting to a single phase, it is recommended to distribute it among the three phases if possible, this to maintain a similar load in each phase and avoid generating electrical problems.

With the defined circuits, the clear technology and the electrical characteristics of the luminaires we can perfectly choose which lighting actuators or dimmers are best suited to our project. However, surely at this point you will find many options that apparently are the same, to define what it is we must go a little more in detail of the actuators.

In the lighting actuators we find very interesting options that give us additional possibilities when programming. These include current detection, ladder function, timings and logic functions. My recommendation is that before choosing the one with the most additional functions, you should analyze if they are really useful for you in the project or not. Here it is also important to keep in mind if the actuators are going to be installed on an electrical panel, in a ceiling or even behind a switch in the electrical box.

As for the buttons or touch screens that control the lighting, among other things, it is important to keep in mind the functions that we will give to each of the circuits. For example, if you have RGB lighting with a KNX controller that allows you to dimmer the channels one by one, because a button to control it would be inappropriate, you would be losing much of the functionality so a better option would be a screen that allows you to display the color palette and select one according to the person's preference.

One last consideration in lighting is that please remember to associate the states. This error is basic but real mind one of the most common even in integrators with experience and certifications, the easy way to realize that you screwed up is that you have to press a switch twice to turn the light on or off. You can check it very easily, you turn on the light of the switch with the switch in question, you turn it off from another place and then press the switch with the switch again, if everything is ok, you should turn it on again but if nothing happens and when you press it a second time if turn it on, then I miss your status.

In general, all loads such as lighting or motors must be distributed as uniformly as possible, in terms of power, in the different phases of the electrical installation. However, it is a good idea to sandwich loads. For example in a certain area having different circuits in different phases, since if damage is generated in any phase we will have lighting that will continue to work. It is also good practice to include some very specific lighting circuits in the UPS load, since in the event of a power outage we will have the option of guiding people to the exits, avoiding possible accidents.

A system closely linked to lighting is the blinds since they are directly linked to the contribution of natural light, so let's move on to the blinds.

Blinds:

In blinds the first thing to analyze is the type of motor to control and define the technology based on that we define what type of controller we will need. Here are several important considerations depending on the type of motors to be controlled. In the case of using 4-wire motors it is very important to specify actuators particularly designed for blinds or at least multifunction with the function of joining two channels for blinds. We can think that it is possible to control this type of motors with two channels of an On / Off actuator and technically if it can, although the cost can be very high. After all, we only need to route the power line to the up or down cable, or disconnect it completely to stop it. The big problem occurs when at the same time we energize the raising and lowering of the motor, since we can ruin it with almost total security. But of course that can be solved by making a logical function on the button that prevents both orders from being sent to the motor at the same time, the problem is that this order could also come from a screen or a cell phone and performing all those software blocks is complicated and very risky. This is why blind actuators are special in that they provide an interlock to prevent this from happening. So use blind or multifunction actuators with the blind option.

In the case of using 12VDC or 24VDC motors, it is essential to ensure that the blind actuator to be used has explicit support for DC motors with polarity change, the recommendation sounds trivial but it is a common error because most of the blind actuators only support AC motors.

Finally, in the case of using RF motors, keep in mind that if there are more than a couple of dozens of motors, they can generate interference between them and cause the system not to work properly. Therefore, it is advisable to use most of these motors with wired control and to supplement with radio frequency in very specific cases or even to use RF to communicate with local boards that operate motors wired.

A good option for blinds is to use flush-type actuators, that is, small actuators that are used for mounting in electrical boxes for example in the electrical box where the blind is connected. This reduces the wiring significantly since you would not have to send the motor cables to the electrical panel and instead you would only need to send the KNX cable to the electrical box where the actuator is located. With all this in mind you can decide what types of actuators you will need and the associated infrastructure. Here it is important to avoid joining large number of motors in the same electrical circuit since they can generate significant electrical noise between them and affect their correct operation. Let's move on to the next subsystem.

HVAC:

The first idea we have is that HVAC systems are difficult to specify, it depends. Let's remember what we previously suggested, lean on the specialists. You as a home automation and automation professional and with the theoretical bases to carry out the integration of HVAC systems have the tools to specify the equipment for the integration in a correct way, so we are going to organize the process.

The first thing is to define what type of HVAC system is needed in the building, as well as the modes, that is, you want to only cool or heat.

We continue to identify different zones for HVAC, that is, which zones require different temperatures. This is basic because it allows you to identify how many thermostats you will have to place in the building. Regarding this, it is important to locate the thermostats in a central place and avoid areas with a direct air flow or that receive direct sunlight continuously as this will affect the temperature measurement.

Remember that the ideal is to use the minimum amount of equipment together, that is, we avoid putting a button next to a thermostat and next to the scene controller you better put a more complete screen or button that includes everything. Even if at the moment you are not going to use any HVAC system, it is best to leave it planned and include at least one device with a thermostat in each zone.

Next we proceed to identify what type of system we are going to operate and based on that we look for whether there is a direct actuator in KNX, such as fan coils, or if we are not going to request that the system have some type of communication protocol. This includes IR control, for example the Split and mini split allow us to control them by IR so we will look for IR emitters in KNX to control the system.

An important detail is to keep in mind the thermal inertia and the hysteresis of the system, we will discuss this with the HVAC specialist and it is important for the implementation that we will see later.

Now let's move on to security systems.

Security:

To correctly specify a security system or integration, the first step is precisely to define whether to integrate an existing security system or to implement one directly in KNX, because there are different modules for security in KNX including alarm panels very complete.

To integrate security systems the most direct option is through dry contacts, that is, the security system delivers dry contacts which are activated in the event of an alarm, however it may also be contacts with some DC potential for this we will use binary inputs to include this in KNX. In case the system allows to use armed or disarmed externally, we will use binary outputs or we can even use modules that have both binary inputs and outputs.

Now if the decision is to implement the security system using KNX, the first thing is to define the vulnerable areas of the building, or those that have a higher risk of intrusion or latent danger. We are going to start by taking up the psychological profile of the client and identifying what is most valuable to him, we must even ask what risks he considers relevant in his building or what is most valuable to protect. We already proceed with this information. The main objective is to protect the people who live there, always. In the second instance we will protect material possessions, but first people. With this in mind we will begin by defining the different areas to protect. We are going to start by protecting the building from intrusion, so with all doors or windows that may have contact with the outside, they must be protected with both an opening sensor and a rupture sensor in the case of glass. The areas associated with these accesses must have a motion sensor dedicated to security and a camera to monitor when the alarm is activated, or we can even use a camera with a motion sensor and later integrate it into the automation system. It is very important that the areas where we have motion sensors for security can be monitored by a camera, since if the alarm is activated in order to protect people, the cause of the alarm activation must be verified by cameras In case of not having a way to see this area remotely we will have to verify it personally and that can be a very risky situation.

Once the building was protected against intrusion, we will proceed to cover internal dangers such as fire. For this we identify the areas where there is direct manipulation of fire and therefore a potential fire risk, here we will locate the smoke sensors. We will also have to cover areas with possible gas leaks with gas sensors, worth the redundancy, and even with air quality sensors. It is important to keep in mind that most of these KNX systems are wired and in fact it is the safest, because being wireless it is easier to generate a carrier blocker that prevents the communication of the sensors making the system vulnerable instead cut the cable inside a tube on the wall is more difficult. Keep in mind that although the sensors ultimately activate an on / off contact, it is not correct to connect them to conventional binary inputs, this because those that are dedicated to security are supervised by the resistance that the module measures, detecting this way if the cable is cut.

With this clear we move on to the next topic, audio and video.

Audio and Video:

Audio and video for a long time was the weak point of KNX, fortunately we currently have direct KNX devices and various gateways that allow us to integrate the most sophisticated entertainment systems.

In this regard, it is best to identify what expectations and preferences the customer has. Based on that we can define if the focus is the video, the audio or both. The process here is very similar to the other subsystems that begins with identifying the zones and defining in each one what requirements we have.

There are wireless HIFI audio systems that we can easily integrate to KNX, these can be associated with different sources and perform multi-room control, that is, define which source will sound in each area at a particular time. On the other hand, if you want to do wiring, you can do it with external audio matrices integrated into the system or directly with KNX audio matrices.

For video it works the same way, we can do multi-room and multi-source control through video matrices. Here the most common is to control the devices by controlling, for which it is necessary to bring the infrared emitter to the device in question. This is very important because we remember that infrared needs line of sight to the device, so we will have to be careful not to leave devices hidden such as blu-ray or television decoders. Let's imagine the scenario in which we leave the decoders and the blueray inside a closed wooden cabinet, later we take the infrared emitters to the interior of the cabinet, because here we can control the devices using KNX without problem but if the client wants to use the Remote control of the device will have problems because it has no line of sight unless you open the cabinet which does not sound very practical. We can solve this type of problem in two ways, the first is by leaving the devices out of closed places and with line of sight, the second option would be to specify a radio frequency remote control that communicates with our KNX system and we replace the normal control of the device with this control.

Another important consideration is that the video and audio system must interact with each other, we will imagine that our client is passionate about soccer and the Champions League, the safest thing is that the final wants to see it on several synchronized televisions and connected to the audio to enjoy with your friends. Let's imagine another scenario where it is a family and they want to see a movie. It is evident that the audio system has to be connected to the video system.

With these considerations we can make the specification of the audio and video system as well as its integration into the automation system.

Energy Measurement and Efficiency:

We now enter energy measurement and efficiency, which is a mandatory objective for the automation of homes and buildings. To know if we are being efficient in energy management or if after the automation project the energy efficiency improved, we have to measure. Measurement is the eye of energy efficiency. Let's remember this:

Automation without measurement is just speculation.

So energy measurement is mandatory and if we can include there measurement of water or gas consumption better. We will start by measuring electricity.

The first thing to define to specify the energy measurement equipment and its integration are the objectives of the measurement in question. That is, what variables we want to measure and monitor in the project. In general, we are going to measure currents, voltages and with this we will have power and therefore consumption. If this is enough, we can use basic energy meters or those intended solely for consumption, bearing in mind that the expected current of the installation is a fundamental parameter for choosing the current transformers that we will use for the measurement or, if it is a direct measurement, depending on the current we can choose one team or another. Now if we require a more complete analysis with power factor, harmonics or alarms, we will have to target equipment type network analyzers. This is particularly useful in real estate since the building must surely have equipment for the compensation of the power factor, but this is usually done at the beginning of the building and as it deals it can generate electrical noise with harmonics that They can damage electronic equipment and even affect energy efficiency.

When we are clear on what type of energy meter we are going to use, we continue with the form and integration, or if we choose to use direct KNX energy meters, it is even easier. In the case that they are external meters, the most recommended is to find that they have a communication card with an open protocol that we will integrate into our KNX system through a Gateway, usually it will be Modbus.

For water or fluid meters, we will have to define what type of liquid we are going to measure and its temperature. Later calculating the maximum flow of each zone we will have what is necessary to specify the meter.

Associated with these equipment, in case of being more than one meter, we will surely need software for energy management. For this, the BMS have specific modules for this purpose. Although we can also do it through scripts if we have the necessary programming knowledge.

So we will now enter to specify the display system that we will use.

Visualization:

For this point we have the design in mind and the specifications of each subsystem as well as its integration. Now the question is how we are going to handle the entire project, since we will have three options mainly.

The first option is that the system is relatively simple and it is sufficient to control it using the different buttons on the project and a maximum of two screens. In this scenario we can use a screen with an application on the cell phone in which we will link the variables to be controlled remotely, we can even do it with a couple of screens. Some of these to control the building outside the local network only require a pairing with the screen and that's it, no more.

The second option is that the project is more complicated or simply more extensive. Imagine for example a one-level apartment of a family with two children that has lighting, blinds, HVAC, security, metering, audio and video. We could use a screen for each area of the apartment and pair with the cell phone, only we would have to have about 8 different screens associated with the cell phone and it can become complex. So at this point we are going to need an additional visualization that brings together all these variables allowing for example to give different permissions depending on the user. If this is the case, we will have to bear in mind to specify the system the number of users that will access the system, the number of different profiles that the system will have, the number of variables to display and the operating systems of the devices that users use. With this of course we can specify the display system to use, the only additional thing to keep in mind is how we are going to control the system remotely. Sure, all display systems work directly on the local network without much work, but it gets interesting when you want to leave the local network. For this, some visualizations have their own associated server that will only ask you to make a registration and you can already control from outside the local network. Some more do not have this option so we have to do it ourselves. To achieve this, we basically have three options: the fixed IP, the VPN and the DDNS.

At this point we already have the specification ready in terms of equipment, integration requirements and important considerations for its installation. This is the most important part in the pre-project stage, however, the start-up of the teams is missing. For this there are also important points to take into account, let's go for it.

24

Efficient commissioning

Commissioning is undoubtedly crucial in any project. Once the system and each of its components have been specified, once installed in their respective place with the pertinent considerations we will put everything to work. So without a doubt we need to pass the design that we have and the functionality that we think to a program that we will later download on the equipments.

For this we will have several considerations in mind, when doing our ETS program. The first thing is to return to our customer profile, the design and the distribution by areas where we had each detailed subsystem and its functionality. We started by defining an addressing structure for our project, here as we mentioned there is no master recipe that applies to all projects as it can change depending on the type of project. However, if there are cross-cutting considerations for all the projects, they will be very useful to you.

We are going to try to group the main subsystems of the

building into the main group addresses, that is, the first numbers of the addresses. In the middle groups the nature of the variables and in the latter the specific areas. There is a reason for this and it is going to make your life a lot easier. Imagine the case where the group "1 / x / y" is the lighting being "x" the type of variable and "y" the circuit in question. For each dimmer circuit we will have, for example, the On / Off variable (1/1 / y), the status variable (1/2 / y), the relative dimmer variable (1/3 / y), the absolute dimmer variable (1/4 / y) and the status variable of the lighting value (1/5 / y). Being "and" the different circuits, for example the room is circuit 1, the dining room will be circuit 2 and the balcony will be circuit 3. So when we see the group address 1/1/1 we know immediately that it is lighting, On / Off and room circuit. If we want to associate the state of the same circuit we know that the group address will be 1/2/1 and if we see for example the group address 1/3/1 associated with a group address On / Off we will be able to see that some variable is has crossed. Of course, as the project grows, this becomes more important because the directions can become confusing and if we do not have a defined structure, the probability of making mistakes will grow exponentially. It is also important to keep in mind that it is a recommendation and for projects with many areas for example we will have to make adjustments to this structure, the important thing is to understand the idea of this example and define a clear group address structure for the project.

Once the structure is defined, we will begin to make the respective associations, always from the most basic to the most complex, keeping in mind to associate the states. Sometimes tend to overlook the basics and start with the most complex, start with scenarios and staircase lighting functions that ultimately complicate the project more than necessary so let's remember we start with the basics and as it works perfectly we are including additional functions and scenarios. This applies to all subsystems, let's think about what are the daily and most trivial functions of each one to start there.

Once we have the basics of each subsist in our program and its additional functions, we proceed to make the systems interact with each other. Only until each subsystem is clear and functional do we proceed to add the interaction between them through, for example, the scenarios. This may sound a bit shocking since it is normal to think about doing the entire program and then download everything, but the truth is that it is more efficient to use the lean startup methodology applied in this area, that is, to iterate continuously and improve incrementally. As the experience increases, it is surely less necessary to iterate and in the end perhaps a single download will be perfect, although the truth is that you always have to iterate little or much, but the fine adjustments are important.

The interaction with the end customer at this point is a factor to have balanced, very little interaction will generate a shock at the end and possibly big changes, but a lot of interaction makes the process slow in the early stages. The most recommendable thing is to define a control matrix for each zone where it is clear how the complete system will work and ask for feedback in finished or almost finished zones.

Let us remember that the scenarios are for the client not for us, that is, the scenarios should make the client's life easier, so it is important to take up the client's profile and the design we have here.

Finally in the start-up it is important to design the best and be prepared for the worst, that is, we will design the best scenarios with the ideal operation but we will test the worst scenario where we know that it could fail. This includes testing simultaneous traffic, permissions, speed of response with streaming, push button operations or improper displays such as multiple quick keystrokes, pressing keys at the same time, etc. Because our task is not to make the system work most of the time, it is to always work even in the worst circumstances.

25

How to finish a project in the estimated time

The time it takes for a project from its design, during its execution and until its final delivery is inversely proportional to the profits and benefits of the project itself. This applies both to you as a KNX partner and to the client, the more time you spend the more expenses you incur and associated problems. The point may come when the project becomes a problem that generates losses for both you and your client, even deteriorating relationships between the two. Thus completing a project in the estimated time is essential.

In order to complete a project in the estimated time at the time of making the quotation, it is essential to estimate the time in the first instance well. It may sound a little obvious but it is not and many overlook it. For example, if the quotation of a project is carried out by the commercial area almost in isolation, it is most likely that they will take too optimistic times, ignoring the details that only an engineer who executes the projects has in mind.

Even sometimes to get along with the client or win a project, very fair times are estimated that in principle make the client be impressed and award the project, but in the medium term these times end by not complying with the consequences that we have already seen and additionally generating the loss of credibility of the company or the project area. So rule number 1 is that a project is quoted in direct relation with the engineers who are going to start it and advised by people who have experience in similar projects, never by the commercial or administrative area in isolation. Clearly, time is a fundamental factor and the client always wants things at a better price, with better quality and in less time, but falling into this can be counterproductive. It is well known that these three things are a balance since if you want to maintain quality by reducing the price, costs will have to go up since you will have to dedicate more human and capital resources to complete the project in less time.

The second great reason why projects tend to take longer than estimated time is due to excessive dependence on other people or companies. For example, to finish a lighting automation project you need lighting, even if you install the drivers and programs until you have the lighting you cannot finish. To solve this, you should try to have the least possible dependence on external factors, both in people, companies or products. The desire to finish quickly can make us think that the faster we start the faster we finish and this is relative, let's take the case that you start working already and end in 6 months because it took 6 months to finish, while if you finish in 7 months but you start in 3 months only took 4 months. Moreover, it may be that really starting in 2 months you continue ending in 6 months, which would be the same delivery date as the first case but with 20% less execution time. Interrupted work intervals cause productivity to drop and your costs to increase, these intervals are directly proportional to dependence on factors external to you. It is much better to start the project a little later if that implies having less dependence on external factors since it reduces the risk significantly. The risk is a fact to take into account when evaluating the costs of a project. This is very clear to the banks. They want to do a risk analysis before setting the interest rate, for example.

The third reason for project delays is staff turnover. The greater your staff turnover, the greater the project execution time, so the ideal is to allocate fixed resources for each project and not move them from there to finish. Here there is nothing more to say simply avoid staff turnover in your projects to the maximum.

The fourth factor to complete the projects in the estimated time is to have processes and protocols defined before the project. Improvisation in a project increases the times, it is better to have the processes defined before starting the execution of the project. With processes I mean all the tasks that must be carried out to finish a project and the way in which they must be executed. This includes how to structure the group addresses in a general way, how to make the control matrices for the programming of the equipment, how to make the flow diagrams of the visualization, how to do the tagging of the cables and how to do the project documentation.

It is important to define specific timelines and goals that lead to the completion of the project, this is essential because it also allows to monitor the progress of the project. The processes must also define the delivery method of the project and the requirements that must be met. For this part it is very helpful to use the agile methodology and the engineering project management plans that can be found in the PMBOK.

The fifth factor that delays the completion of a project are the additional elements that arise in its execution. Here the recommended is to define a clear scope of the project and handle the additional ones as different mini-projects, this allows to finish the first contract without increasing the execution time. The ideal is to carry out partial deliveries according to the established scope and in case of not being able to handle the additional ones as contracts, the calendar should be explicitly modified, also adding the costs that this implies.

So summarizing to finish a project in the estimated time we start estimating the time well including a space for contingencies supported directly in the technical area, minimize external dependency, have previously established processes and clear schedules with clear partial deliveries, follow up newspaper and seek to handle the extras as separate contracts.

26

Diagnosis and problem solving

A good diagnosis and problem solving in a KNX project can differentiate a good integrator from a not so good one. In order to solve the problems of a project, a deep and detailed diagnosis is essential. Here we have to identify the cause of the problem in an absolutely objective way, doubting everything and making sure that everything works correctly.

We will start by reviewing the main electrical grid since many times the apparent errors of the automation system turn out to be problems of the electrical grid as such. This includes checking the connections of the KNX bus and the automation board, the efficient way to do this is to start with the equipment close to where the problem occurs in the installation.

Assuming that the electrical installation is fine and the physical connection of the equipment is also correct, we proceed to make a diagnosis through the ETS. Let's start by checking the physical addresses using the "Diagnostics" and "Individual Addresses" module. Here we will have three options.

The first option "Programming Mode" allows us to identify which devices are in programming mode. This is particularly useful when for example we are trying to program a device and it indicates that more than one device is in programming mode so we can see specifically which device. additional is in programming mode quickly.

The second option "Verify Individual Address" allows, as its name indicates, to see if a particular equipment has this physical address, particularly useful when we want to verify that we have carried out the programming of the physical addresses correctly or even if some equipment was preprogrammed .

The third option "Line Scan" allows us to verify that individual addresses are physically connected to a particular line. This is particularly useful if we have a bus loss somewhere or equipment that apparently does not respond, if we do not see that they are connected after the line scan, it may be a connector problem and indeed the equipment is isolated from the communication bus preventing its correct operation.

Once the physical addresses and the devices connected to the bus have been verified, as well as the electrical installation as such, we will verify that the program download and the group addresses associated with each equipment are in its memory. For this we use the option "Device information" and "Device information (with associations between groups and objects)".

Next we will review the design again on paper and the topology, making sure that we are respecting the theoretical foundations and that we are not exceeding the limits of the

KNX structure. And we will sit down to think what could cause the problem, about what situations arise and we will document this. Analyzing what stimuli within those particular situations can trigger the error or anomalous behavior. We will make different hypotheses before continuing and recreate the error as accurately as possible, that is, we will make the system fail completely intentionally. This step is very important and although it seems a bit obvious most of them skip it and begin to review the telegrams trying to guess what happens without any clue, Henry Ford said it well:

"Thinking is the most difficult job, that's why so few people do it. "

With the clear situations where the error occurs and possible causes of it, as well as hypotheses of what is happening, we are going to analyze the telegrams that are flowing on the bus to better understand what is happening. Here the objective for now is not to solve anything, it is to understand what happens and why it happens, and later if to look for a solution.

With the group monitor it is usually sufficient to identify the problem and solve it. So we will enter the group monitor and here the objective is simple, we will recreate the situation in which the error is generated, we will analyze the telegrams that flow on the bus. If we identify unwanted telegrams we can go to the team that is sending them and review their parameters to see what is happening. In case all the telegrams are apparently correct and the values in the telegrams are as indicated, we will go to the associated group addresses to check the flags on the group objects. The ETS provides the option to record telegrams in case they are too many and it is very difficult to do the analysis in real time.

Finally we have the option "Verify Project" where we will obtain information on the devices, topology, group addresses and product data. With these tools we can identify the problem and then search for its solution.

The ETS is a fairly interactive and intuitive tool so I highly recommend installing the demo version and exploring it to familiarize yourself with the tool. The KNX association has a simulator so there is no need to buy KNX equipment to learn, we can do it with the demo license in the comfort of our home and without buying additional equipment.

27

After-sales service

The after-sales service is one of the most important issues and is one of the most dangerous ones. It is very true that the customer is the most important and the after-sales service must be timely, competent and pleasant. It is also very true that if the limits are not very well defined between you as an integrator and the client, regarding the scope after the delivery of the project, then it can go from being after-sales service to after-sales slavery.

Once the limits are defined and the scope is very clear, the perfect project must be delivered with all the associated documentation, and an annual maintenance plan must be proposed, for example where the client invests and you can sustainably maintain yourself. Otherwise you can see yourself immersed is a lot of grateful but low-paid work that ultimately ends up harming the client, if your work is not properly paid your business model is not profitable and if it is not profitable the quality will inevitably drop even forcing you to withdraw from the game which ends up damaging the customer and of course you. So defining the scope and paying for the maintenance associated with the system is a good idea for both you and your client.

Leaving the above very clear, we proceed to analyze the after-sales service. Here you have to keep in mind what type of client you are serving. In the event that your client is institutional, or is a company, then it is easier because problems in principle can arise during office hours and it is easier to foresee the necessary infrastructure to react to this type of eventuality, even personally. Now if your client is domestic or residential, the after-sales service changes completely since the probability that eventualities occur outside normal working hours is absolutely high, it is to be expected since most people will be working during the day and when they get home to use the system it will be when eventualities may arise. For this it is very important to define timely reaction systems where they can even be automated. If you listened well, many of the problems that arise in home and building automation can be solved by carrying out a specific protocol, so you could foresee some of the eventualities that will arise and how to react to them. Therefore, you could create a customer service system that takes the user through a defined protocol to solve problems and if it is definitely a problem that requires the attention of staff, you will have to react in a timely manner, the important thing here is to define processes and protocols can ease the burden and automatically fix most problems.

This part is important for your business model as it allows you to change an occasional client with a specific value associated with the project to a client with a "life value" that offers better service and more "engagement" you manage to generate will be more profitable.

Home and building automation is the present, until here you have the tools to deal with what exists so far and start your journey as an integrator. Now we will see what to keep in mind because it is the future and very soon it will be the present so the faster you start preparing the more ready you will be when opportunities arise.

The following part applies to integrators, developers, manufacturers, training centers and administrative personnel. So let's go for a walk in the future.

28

How to increase the profits of your business

The objective of a business is to obtain profits, if the impact is important and to make a better world it is fundamental as a purpose of life but, realistically, the objective of our business is to have profits and those profits are those that allow executing a mission in a sustainable way. A business that does not have profits is not a business because of the positive impact it generates in the world, so here we will see some ideas so that you can reflect on your particular business model and increase the profits of your business.

There are mainly two ways to increase profits, understood as net profit, which are to increase income or decrease costs.

Let's start with the costs, because if you have a client or a hundred you can always optimize costs and resources. Normally, a KNX integrating company will have three areas associated with its operation: commercial, administrative, and technical. Beware that here everyone tends to think that their area is the most important, a commercial tends to think that their area is the most important of the company because it is the one that sells and brings the money, an administrative officer will think that their area is the most important. important because it is the one that structures the business and manages the resources making the entire company viable, a technician will think that his area is the most important because he is in charge of making projects come true and complying with customer requirements, making the company really do projects. But the truth is that all three are important, it's like a three-legged table. If you remove any one, it may hold a little but it will fall. In each of the three areas there are different options to optimize costs.

• Optimize costs in the commercial area: The commercial area has different marketing strategies and customer acquisition associated, its ultimate goal is to sell and bring new customers to the company. Here it is essential to analyze the customer profile that we did previously since this will allow efforts to be focused correctly, which will translate into greater customer conversion. Analyzing the developed customer profile also allowed selecting the channels that best fit the segment we want to capture. Here we can apply the 80/20 rule that tells us that 80% of the results come from 20% of the activities, so focusing on the customer profile allows us to locate that 20% more easily. Additionally, including a programmer engineer in the commercial area will optimize costs, since most of the tasks of marketing campaigns can be automated to run autonomously. Additionally, in the commercial area looking for a positioning approach instead of looking for leads will optimize costs in the medium and long term, let's see it this way looking for leads is like going fishing with a spear that requires great effort and the conversion rate is not so high, on the other hand to do positioning is to fish with a network where the effort is less and the conversion rate is higher since the clients that arrive want to do business. Finally, it is very useful to rely on digital platforms, especially those that are undervalued in your specific market niche.

• Optimize costs in the technical area: The technical area has its greatest opportunity to optimize costs in the design and optimization of projects. Having defined, documented and

ready-to-apply processes in a project allows, in the first instance, to reduce time in project design and budget, second, allows to reduce time in the execution of projects, third, allows to increase quality and predictably meet technical requirements clearly. defined. For this it is highly recommended to use engineering project management and agile methodologies.

• Optimize costs in the administrative area: The administrative area has interesting opportunities for improvement to reduce costs. The first is to reduce the inventory of physical equipment that we have, despite the fact that we tend to think that the inventory in projects reduces time and allows us to win more projects. The truth is that most of the time that becomes equipment stored in a cellar that end up being unused. So reducing physical inventory is essential, this is compensated by having the times of the supplier channels very clear so that they can be included in the project management schedules mentioned above. The second opportunity in the administrative area is to reduce the variety of equipment so that it is easier to standardize processes, join projects to make larger orders, and thus obtain better conditions with suppliers. Look at it this way, it is much more advantageous for you to negotiate an order with a supplier of 100 units of 3 teams in a single order than to be ordering 2 units of 10 different teams every month. The third opportunity is found in trying to pass fixed costs to variable costs, understanding as fixed costs what you have to always pay regardless of whether you sell or not, while variable costs are those that depend on your sales. Fixed costs would be for example the rent of your office, payroll and internet services while your variable costs would be commissions, sales taxes and the people you hire for specific projects. For example, if your commercials have 100% fixed salaries are an absolutely fixed cost, an option would be to manage a basic plus a commission associated with sales, this, in addition to converting a fixed cost into a variable, will have a positive impact on the performance of the personnel. A fourth option is to support administrative processes such as inventories, accounting and management in digital tools that there are several in the market or even scripts can be scheduled to help you optimize time and reduce costs at the end of the day.

Now to increase income the immediate reaction is increase sales or increase prices. To increase sales, that is, to have more clients or projects, we rely on the value proposition and the business model canvas in Chapter 21. While to increase prices, a direct option is to deliver more value to the client, solving a pain or a problem. greater concern, or generating greater profits or joys. Let's remember that all the decisions we make as people are reduced to fleeing pain or pursuing pleasure, so if we want to increase the price we have to focus on one of these two options.

However, to increase income there are four fundamental ways.

First if you have 100 clients the objective is to increase the number of clients, we have already talked about that. Second is to sell more to each client, let's take as an example companies like McDonald's where you usually go for a hamburger but you come out with an ice cream, potatoes and enlarged soda, this means that despite being the same client the purchase is greater and therefore the income for the company increases, this way is very interesting because it is much easier to sell more to a client than to get a new client. The third way is to increase the frequency of transactions of each client, the objective is that if a client does an automation project with KNX every 6 months then we will look for him to do a project every 3 months doubling the income, to achieve this we can use offers or close accompaniment to each client to identify what opportunities are missing. The fourth and final way is to increase the prices we have already analyzed, focusing on the supply of value.

IV

DEVELOPMENT

First of all, I congratulate you for reaching this point. If you have put into practice what you have seen in this book and have done a couple of integration projects, I am sure you have a very good level. Just for that we are going to move to the next level. I am going to show you how to create your own server to control KNX systems. The goal of this chapter is to open your mind.

"The first step is to establish that something is possible, then it will probably happen" Steve Jobs.

29

KNX test system

To start we need a basic KNX system to handle. We already dedicate good time to KNX systems, so here we will keep this step simple but well done. So we are going to use four basic equipments: a KNX source with auxiliary power supply, a minimum two-channel On / Off actuator, a KNX button and a KNX / IP router.

As for complementary materials we are going to need an electrical connection plug, a 40 cm omega rail, an electric breaker, a rail fuse holder with an ultra-fast fuse, 50 cm of 4-wire KNX cable, 1 meter of gauge cable 16 awg, 2 control terminals and two LED bullets. You can dispense with a large part of these materials and do the assembly with insulating tape and on the carpet in the room but well we are going to do it well, so assuming that you already have the materials we proceed.

We begin by placing on the rail the breaker, the fuse holder with the fuse, two control terminals, the KNX source, the KNX actuator and the KNX / IP router in that order. We are going to connect pin with the control cable to the top of the breaker and to the top of the first control terminal, identifying the tip of the breaker as "L" and the tip of the terminal as "N" both in the pin and in the lathe and the breaker. We continue by connecting with a cable the breaker output to the top of the fuse holder and the bottom of the fuse holder to the top of the second terminal making up this terminal as "L". Now we will connect the lower part of the first terminal to input "N" of the source and the lower part of the second terminal "L" to input "L" of the source with a control cable. We continue connecting the input of the two channels of the On / Off actuator between them and we will connect these with another cable to the bottom of the "L" terminal. Now we take the LED bullets and connect the neutrals to the bottom of the "N" terminal and each of the remaining two cables to the outputs of the two channels of the On / Off actuator.

At this point we already have the force connection of our test system ready, we go to control. We take the KNX cable and connect the red / black cables from the source to the actuator, after the actuator we connect the KNX button with the red / black cables. Finally we connect the IP interface to the source using both the red / black communication bus and the additional yellow / white power supply.

Great!!, we already have the armed system so we are going to program it. We start by downloading and installing the ETS software, with the demo license it is enough for us so there is no need to worry.

In our project we are going to create a line 1 and add the equipment in the following order:
- KNX / IP router: Physical address 1.1.0
- KNX actuator: Physical address 1.1.1
- KNX button: Physical address 1.1 .2

In the KNX / IP router we are going to assign it a fixed IP address that is 192.168.1.100.

In the KNX actuator, in case we are going to select that the channels work independently and type On / Off, we are going to enable the sending of states in different objects and we will enable the sending of the state after change, we will also configure two scenes. For scene 1 we assign channel 1 the value On and channel 2 the value Off, for scene 2 we assign channel 1 the value Off and channel 2 the value On.

In the KNX button we are going to use only 1 Gang, that is, a double key. For the first gang then we select that both buttons work independently with toggle type or switching. Here we are also going to enable the reception of state in a separate object.

Now we are going to create two main group addresses that are 1-Illumination and 2- Scenes, within illumination we will create two intermediate addresses that will be 1/1 On-Off and 1/2 State, within each of them we will create two addresses of group being x / x / 1 LED Ceiling and x / x / 2 LED Wall. Finally within the main address 2-Scenes we will create an intermediate group called 2/1 General and within this a group address 2/1/1 Lighting so the group addresses should finally be as follows:

- (1) Lighting - (1/1) On Off => 1/1/1 LED Ceiling, 1/1/2 LED Wall
- (1) Lighting - (1/2) Status => 1/2 / 1 LED Ceiling, 1/2/2 LED Wall
- (2) Scenes - (2/2) General => 2/1/1 Lighting

We already have the parameterized equipment and the group addresses created so we are going to make the associations. We take the switching objects or On / Off of the two channels of the actuator and associate them to the addresses 1/1/1 and 1/1/2 respectively. Now we take the status objects of the two channels of the actuator and associate them to addresses 1/2/1 and 1/2/2 respectively. We proceed by associating the scenes object of the actuator to the address 2/1/1.

Now we go with the KNX button. We are going to take the switching objects from the two pushbutton keys and associate them to the group addresses 1/1/1 and 1/1/2 respectively. We continue taking the status objects from the two pushbutton keys and associating them with group addresses 1/2/1 and 1/2/2 respectively.

With this we are going to do a total programming of the equipment and we are ready to make our own application that will control the system. So let's go for it!!.

30

Creating your own KNX server is possible

We will go through it step by step so that you can create your own code, however I am aware that sometimes it can be a bit overwhelming and personally I think that the examples are important so You will have the complete codes for you to use in your process with total freedom so if you prefer to see all the code and then if you read this part later you will have them here https://github.com/angelioe/lossecretosdeknx. You are about to go to the next level and see that it is possible to not only use technology but also create it.

To achieve this we are going to make an incredible architecture for the advantages it gives you and what you will learn in its development, it is also fully scalable and if you wish it can be the starting point of your industry. This architecture is made up of two different applets and a database. We will call the first subprogram KNX Deamon, this will take care of entering the KNX system monitoring the telegrams and executing actions when necessary. The second subprogram will be called Flask App, this is responsible for generating the web server where we will see the KNX variables and we can control the system. The database is an intermediary between Flask App and KNX Deamon, that is, the communication between these two programs will be through the database both to read KNX information and to generate actions in the system.

Database:

The set is made up of two different databases with delimited functions. The first table is the instructions table, this is in charge of storing the instructions that the system must execute as well as their status, that is, if they have already been executed or not. The second table, called data, is in charge of storing the history of the actions executed in the system and the state variables of our KNX system.

KNX Deamon:

We are going to use the python xKNX library monitoring the KNX bus telegrams and injecting them to control the system. To carry out the integration in terms of KNX infrastructure, it is necessary for the KNX system to have a KNX / IP interface or a KNX / IP router, additionally it is essential to know in detail the group addresses used in the ETS and the type of data in each one of them.

Now on the computer it is necessary to install some previous libraries that are detailed below, as well as their function:
 • netiface 0.10.7: Its main function is to list network inter-

faces on the local machine, as well as network addresses and even preserves portability.

• pyyaml 3.13: Its function is to analyze and issue YAML files, used mainly for human reading and interaction with scripting languages.

Programming should not be linked to the programming language, it sounds a bit strange but the point is that a good programmer is a good programmer for his way of thinking and structuring the solution to a problem independent of the programming language. If it is true that each language has its peculiarities and it is necessary to understand them thoroughly to optimize the program, but here that is not our objective and then if you want to optimize the program you will have time for it. Therefore, the most important thing before starting to think about the syntax is to define the structure and flow of the program, and then we will see how we write it.

Let us remember again that this program will be running in the background and will not be manipulated by the user, the web server or Flask App will write the commands in a database and our KNX Deamon program will be pending from the database to execute the pertinent actions. Likewise, if the KNX Deamon finds a change in any variable, what it will do is write it to the database and the web server will take it from the database.

To do the complete exercise we are going to generate a group monitor to see the telegrams in real time, since you are convinced that it is possible to create technology so here we are going to do it.

Once the program starts you have to connect to the database, then you will connect to the KNX system using the IP defined in our KNX / IP Router. Once connected to the system, we will have to organize two tasks that will be apparently simultaneous, although they are actually executed asynchronously, that is, they are interspersed very quickly and depending on the pending processes. These two tasks are reading the database to see if the web server sent a new command that the KNX Deamon should execute or if there is a new telegram on the KNX bus that should be displayed on our monitor, we will see one by one.

In case a new telegram is received on the KNX bus we will have to check that this telegram is not already stored in our database, only in case it is not stored we will proceed to save it in the database. And we will be pending of new telegrams again.

On the other hand, if we read a new order in the database that is pending, we will check if it is for the KNX system, in case it is, we will execute the action of inserting said telegram on the KNX bus. Once the telegram has been inserted, we are going to modify the status of the instruction from "pending" (p) to "done" (d), thus registering that the instruction has already been executed. We could only delete the executed instruction from the database, but by doing this state system we are going to leave a history of the actions which is very useful for traceability and diagnosis.

We see below the flow diagram of our KNX Deamon program.

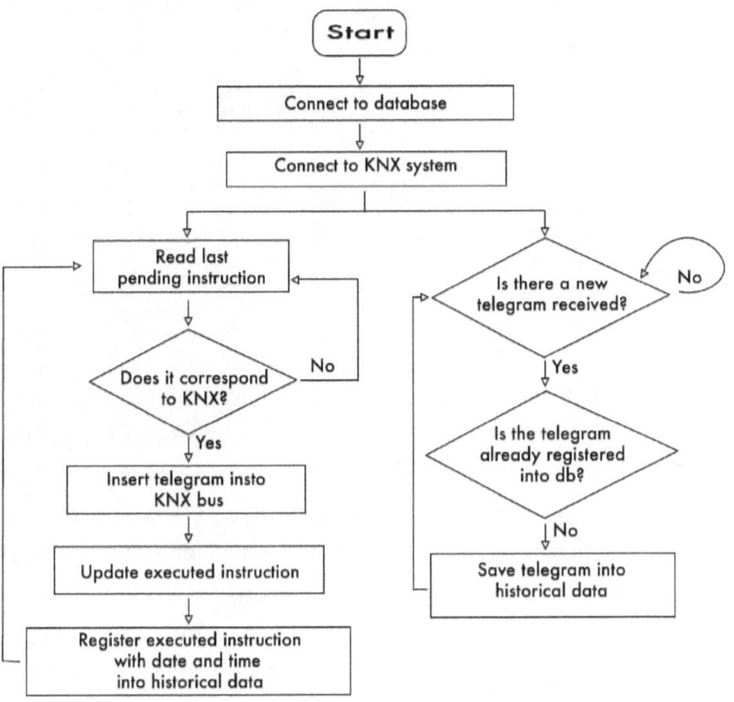

Well let's start with the code, for this we need to do four functions.

The first will be to identify whether we have internet access on our local network or not. The idea is to make a request to a page that is usually online, we will use Google's. If the answer is state 200, it means that it was successful, if it is between 400 and 600, it means that it was not successful, therefore we would not have internet access. We will call this function internet_on().

```python
def internet_on():
try:
http = urllib3.PoolManager(timeout=3.0)
url = 'http://www.google.com'
response = http.request('GET', url)
if response.status == 200 : return True
if response.status >= 400 or response.status <=600:
    return False
except urllib3.URLError as err:
return False
```

The second function that we are going to use is in charge of monitoring the telegrams and the management of telegrams received on the KNX bus. For this we are going to start by connecting to the database since there we will store the received telegrams and then be read by the web server. Next we are going to take the received telegram and convert it into a chain tipe "string". We continue taking the hour and the day of the week concatenating them separated by a space. Remember that the telegram is a packet with a lot of information so the next step is to divide that telegram into useful information such as group address, DPT, value, telegram type and address. With the segmented information, we continue to save it in the database in case it is not repeated. For this we first create a cursor of the database connection and with that cursor we are going to take from the database the last record that matches "knx", the group address of the telegram that we want to save and the DPT. If we obtain information from this, it means that there is already a stored telegram with that group address and DPT, then we will proceed to compare whether the value stored is the same that we are going to save, just in case be different we proceed to save the telegram. On the other hand, if we do not obtain information from the last telegram with that group address and DPT, we are sure that the telegram is not repeated, so we proceed to save it. We finish by saving the changes made in the database and closing the connection so that it can be used later.

```python
async def telegram_received_cb(telegram):
connt =
sqlite3.connect('/angel/FlaskApp/hopeitworks.db')
telegram_str=str(telegram)
date_time=str(datetime.datetime.now())+"
"+str(datetime.datetime.today().weekday())
group_address,DPT,value,telegram_type,direction=
        split_telegram(telegram_str)
c = connt.cursor()
c.execute("SELECT VALUE FROM DATA WHERE CONTROLLER=?
AND DEVICE=? AND FUNCTION=? ORDER BY ID DESC LIMIT
1",("knx",group_address,DPT))
all_rows = c.fetchall()
if (str(all_rows) != "[]"):
if (str(all_rows)[3:-4]!=value):
print(telegram_str)
c.execute("INSERT INTO DATA
(DATE_TIME,CONTROLLER,DEVICE,FUNCTION,VALUE)\
 VALUES
 (?,?,?,?,?)",[date_time,"knx",group_address,DPT,value]);
else:
c.execute("INSERT INTO DATA
(DATE_TIME,CONTROLLER,DEVICE,FUNCTION,VALUE)\
 VALUES
 (?,?,?,?,?)",[date_time,"knx",group_address,DPT,value]);
connt.commit()
connt.close()
return True
```

The third function that we are going to need is the one that divides the entire string of the telegram into useful information that we can store in the database. This is simple since we will have the information separated with "," therefore it is enough to create the variables and apply the Python split function for strings.

```python
def split_telegram_ins(telegram):
telegram_str=str(telegram)
id,device,function,value = telegram_str.split(",")
id = id[2:len(id)]
device = device[2:-1]
function = function[2:-1]
value = value[2:-3]
return(id,device,function,value)
```

The fourth function is the main one and is responsible for managing all the others, as well as reading the database constant to see what orders the web server sends to be executed in KNX. We start by creating the xknx instance and defining that we are going to use the function "telegram_received_cb" for the management of incoming telegrams. We continue by starting the xknx instance asynchronously and sending the value 1 in the group address 0/0/1 just to visually check that it is working, this line could be omitted without any problem. Then we begin the cycle in which our KNX web server will be, which begins by connecting to the base of data and creating the corresponding cursor. Then we take from the instruction table the last stored order for KNX that is in a pending state and if this is not null we separate the string in the different parts of the telegram, we continue checking that the function is of the binary type since we are only working with On / Off variables and if so, we put the telegram in the output queue to be entered on the KNX bus. As the action is already executed then we update the status of the order to executed (d). To finish we have to save the change in the variable executed by the command so we proceed in the same way as in the function "telegram_received_cb". We end up saving the changes executed in the database and closing the connections, here we must be careful that the xknx.stop () is outside the while loop.

```python
async def monitor(address_filters,db):
xknx = XKNX()
xknx.telegram_queue.register_telegram_received_cb(
telegram_received_cb, address_filters)
await xknx.start()
await
xknx.telegram_queue.process_telegram_outgoing(
Telegram(GroupAddress('0/0/1'), payload=DPTBinary(1)))
while True:
conn = sqlite3.connect(db)
c = conn.cursor()
c.execute('SELECT {id},{coi1},{coi2},{coi3} FROM {tn}
WHERE {cn}="knx" AND {cs}="p" ORDER BY {ord} DESC
LIMIT 1'.\
format(id='ID',coi1='DEVICE', coi2='FUNCTION',
coi3='VALUE', tn='INSTRUCTIONS', cn='CONTROLLER',
cs='STATE', ord='ID'))
all_rows = c.fetchall()
if (str(all_rows) != "[]"):
id,device,function,value =
split_telegram_ins(all_rows)
if (function == "bin"):
await xknx.telegram_queue.process_telegram_outgoing(
Telegram(GroupAddress(device),
payload=DPTBinary(int(value))))
c.execute('UPDATE {tn} SET {cs}="d" WHERE {ide}=id'.\
```

```python
format(tn='INSTRUCTIONS', cs='STATE', ide='ID'))
date_time=str(datetime.datetime.now())+"
"+str(datetime.datetime.today().weekday())
c.execute("SELECT VALUE FROM DATA WHERE CONTROLLER=?
AND DEVICE=? AND FUNCTION=? ORDER BY ID DESC LIMIT
1",("knx",device,function))
all_rows = c.fetchall()
if (str(all_rows) != "[]"):
if (str(all_rows)[3:-4]!=value):
conn.execute("INSERT INTO DATA
(DATE_TIME,CONTROLLER,DEVICE,FUNCTION,VALUE)\
 VALUES
 (?,?,?,?,?)",[date_time,"knx",device,function,value]);
else:
conn.execute("INSERT INTO DATA
(DATE_TIME,CONTROLLER,DEVICE,FUNCTION,VALUE)\
 VALUES
 (?,?,?,?,?)",[date_time,"knx",device,function,value]);
conn.commit()

conn.close()
await asyncio.sleep(0.1)
await xknx.stop()
```

To finish our code what we will do is generate our "main". We start by making the connection to the database and creating the tables that we are going to use, in case they are already created we simply create a variable in case we want to save the record. We continue creating the asynchronous loop since our functions are asynchronous, we will use this method, we continue checking the internet connection and saving this in a boolean variable. Then we run our main function which is "monitor" in the asynchronous loop that we create until it is complete and finally we close it.

```python
if __name__ == "__main__":

conn =
sqlite3.connect('/angel/FlaskApp/hopeitworks.db')
db = '/angel/FlaskApp/hopeitworks.db'
try:
conn.execute('''CREATE TABLE INSTRUCTIONS
 (ID INTEGER PRIMARY KEY AUTOINCREMENT,
 DATE_TIME TEXT NOT NULL,
 CONTROLLER TEXT NOT NULL,
 DEVICE TEXT NOT NULL,
 FUNCTION TEXT NOT NULL,
 VALUE TEXT NOT NULL,
 STATE TEXT NOT NULL);''')
conn.execute('''CREATE TABLE DATA
 (ID INTEGER PRIMARY KEY AUTOINCREMENT,
 DATE_TIME TEXT NOT NULL,
 CONTROLLER TEXT NOT NULL,
 DEVICE TEXT NOT NULL,
 FUNCTION TEXT NOT NULL,
 VALUE TEXT NOT NULL);''')
except:
e = "Database already created"

conn.close()
loop = asyncio.get_event_loop()
if internet_on(): internet_bool = True
else: internet_bool = False
loop.run_until_complete(monitor(None,db))
loop.close()
```

At this point you may be lost, what I usually do is put things
to work and then understand how it works, mixing theory
with practice. So I encourage you to test the code only in a
terminal and go including commands in the database using
"sqlite3" in another terminal window. With this really working

we have the vast majority of the work done, now we will go to the graphical user interface.

Flask App:

As a framework we are going to use Flask for the advantages it presents for this development:

• It is an Open Source system and is covered under a BSD license.

• Good documentation, GitHub code and mailing list.

• Its agility and practicality allow to develop simple applications quickly and efficiently.

In this project, the architecture is distributed in different programs, which allows maintaining practicality in each one, including the web application.

• It has an integrated debugger which allows to make an efficient diagnosis of the faults that appear in the development process.

And the design of the web page we will use CSS with dynamic design to provide a natural and comfortable user experience.

Once each control is pressed by the user, the server generates the instruction in its respective format and writes it in the instructions table of the database with a "pending" status, later the "deamon" type program responsible for the system in which the device to be controlled is in charge of executing it as previously specified.

The server communicates in real time with web clients through

WebSockets, allowing full duplex communication between client and server, therefore, anyone can send information to the other through an established connection. For this we will use the Flask-socketio extension, active communication via web-socket will be implemented through a thread in "background".

For our Flash server to work it is simple we only need a python file that will create the application to manage the server and a couple of templates that will be the web pages that we will use as the user interface on our server, we will only use two pages one where we will have control of the two lights and one with the group monitor for telegrams, thus covering both control and monitoring. The start page with the control of the lights we will call "inicio.html" and the monitor "friki.html", if we are going to also use html, CSS and javascript but calm it is not necessary to know these languages for now.

Flask App (app.py):

Let's start with the Flask application. The first thing, after importing the libraries, of course, is to create the "app" application and the "socketio" communication. It is also important to define a thread that we will use to send the KNX bus information in real time:

```
app = Flask(__name__)
app.config['DEBUG'] = True
app.config['SECRET_KEY'] = 'super-secret'
socketio = SocketIO(app, async_mode=None)
thread = None
thread_lock = Lock()
```

Now we are going to define a couple of functions that we are going to use. We will start with the function in charge of sending the information that is circulating on the bus in real time and in the background. We start by defining a delay time of 0.3s to avoid saturating the server, we continue connecting to the database, generating a cursor on it. Then we take the last 10 records from the instruction table in the database and put them together in a list. This list is encoded as JSON to be sent to the client and we do the same process for the last 10 records of the data table of the database. Finally we send the two lists encoded through socketio.emit:

```python
def background_thread():
    """Example of how to send server generated events
        to clients."""
    while True:
        socketio.sleep(0.3)
        xconn = sqlite3.connect('./FlaskApp/
            hopeitworks.db')
        xconn_c = xconn.cursor()
        xsys = xconn_c.execute("SELECT * FROM
            INSTRUCTIONS ORDER BY ID DESC LIMIT 10")
        system=[]
        for rows in xsys: system.append({'ids': rows
            [0], 'dates': rows[1], 'controllers': rows
            [2], 'devices': rows[3], 'functions': rows
            [4], 'values': rows[5], 'states': rows
            [6]})
        xsys_json = json.dumps({'logs':system})
        xtel = xconn_c.execute("SELECT * FROM DATA
            ORDER BY ID DESC LIMIT 10")
        telegram=[]
        for row in xtel: telegram.append({'idd': row
            [0], 'dated': row[1], 'controllerd': row
```

```
        [2], 'deviced': row[3], 'functiond': row
        [4], 'valued': row[5]})
    xtel_json = json.dumps({'telegrams':telegram})
    socketio.emit('my_response',{'data':xtel_json,
        'sdata':xsys_json})
    xconn.close()
```

Now we need to do two functions for each of the pages that we are going to use. A function that will send the client to the corresponding HTML template and another function that manages the orders sent by the client. In the main path we define a global thread, launch our background function and return the home HTML template.

```
@app.route("/")
def es():
    global thread
    with thread_lock:
        if thread is None: thread = socketio.
            start_background_task(target=
            background_thread)
    return render_template('home.html',  async_mode=
        socketio.async_mode)
```

The following function for the main route will manage the sent orders. Initially we are going to do a conditional where if the information sent by the client is "friki" then we redirect the client to the monitor template called "friki". In case the conditional is fulfilled, the flow continues and means that the client sent a KNX order, therefore we are going to divide the information sent into controller, devices, function and value. With the useful information already segmented we are going to connect to the database, take the current time and insert

the command in the instruction table for the KNX Deamon
program to execute it. We finish by closing the connections
and making a null return:

```python
@app.route('/', methods=['POST'])
def es_post():
    if (request.form['action']=='friki'):
        return redirect(url_for('friki'))
    controller,device,function,value =
    str(request.form['action']).split("-")
    conn =
    sqlite3.connect('./FlaskApp/hopeitworks.db')
    try:
        conn.execute('''CREATE TABLE INSTRUCTIONS
                (ID INTEGER PRIMARY KEY
                AUTOINCREMENT,
                DATE_TIME TEXT NOT NULL,
                CONTROLLER TEXT NOT NULL,
                DEVICE TEXT NOT NULL,
                FUNCTION TEXT NOT NULL,
                VALUE TEXT NOT NULL,
                STATE TEXT NOT NULL);''')
        conn.execute('''CREATE TABLE DATA
                (ID INTEGER PRIMARY KEY
                AUTOINCREMENT,
                DATE_TIME TEXT NOT NULL,
                CONTROLLER TEXT NOT NULL,
                DEVICE TEXT NOT NULL,
                FUNCTION TEXT NOT NULL,
                VALUE TEXT NO T NULL);''')
    except:
        e = "Database already created"
    date_time=str(datetime.datetime.now())+"
    "+str(datetime.datetime.today().weekday())
    conn.execute("INSERT INTO INSTRUCTIONS
    (DATE_TIME,CONTROLLER,DEVICE,FUNCTION,VALUE,STATE)\
```

```
                        VALUES (?,?,?,?,?,?)",[
date_time,controller,device,function,value,"p"]);
    conn.commit()
    conn.close()
    return ('', 204)
```

We apply the same logic for the function of the route / friki:

```
@app.route('/friki')
def friki():
    global thread
    with thread_lock:
        if thread is None: thread =
        socketio.start_background_task(
target=background_thread)
    return render_template('friki.html',
    async_mode=socketio.async_mode)
```

And since in the monitor we are only going to supervise the telegrams then we will have no control requirements, only to return to the main page :

```
@app.route('/friki', methods=['POST'])
def friki_post():
    if (request.form['action']=='home'):
        return redirect(url_for('en'))
    return ('', 204)
```

And to finish we have the main function of our application that in effect launches the application on port 5000:

```
if __name__ == "__main__":
    conn =
    sqlite3.connect('./FlaskApp/hopeitworks.db')
```

```python
try:
    conn.execute('''CREATE TABLE INSTRUCTIONS
            (ID INTEGER PRIMARY KEY
            AUTOINCREMENT,
            DATE_TIME TEXT NOT NULL,
            CONTROLLER TEXT NOT NULL,
            DEVICE TEXT NOT NULL,
            FUNCTION TEXT NOT NULL,
            VALUE TEXT NOT NULL,
            STATE TEXT NOT NULL);''')
    conn.execute('''CREATE TABLE DATA
            (ID INTEGER PRIMARY KEY
            AUTOINCREMENT,
            DATE_TIME TEXT NOT NULL,
            CONTROLLER TEXT NOT NULL,
            DEVICE TEXT NOT NULL,
            FUNCTION TEXT NOT NULL,
            VALUE TEXT NOT NULL);''')
except:
    e = "Database already created"
conn.close()
socketio.run(app,host='0.0.0.0',port=5000)
```

With our complete application we will go to the two HTML
templates.

Starter template (inicio.html):

With this book I provide you with the complete templates in
the repository at the beginning of this chapter and since these
are extensive and most of the code is aesthetic we are only
going to see the purely functional parts of the templates to
then run the codes, make modifications or make your display
from scratch according to your preference.

The start template has the function of sending orders to the KNX system, we will see an example of a button with pop up to do this. We start by making a container to destroy our button for the ceiling lighting. Next we create a title and a separator using w3 styles. Then we create our button and define that by clicking it sends us to the script "openForm", which is responsible for making the pop up visible where we will see the buttons for turning the light on and off. So we proceed to create the pop up with id "lv_lso_popup", note that it is the same id that we are passing in the button created just before since this will be the one that we will enable with our script. Inside this pop up we create a form associated with the "POST" method so that it sends the information to the function created in the Flask application associated with the same "POST" method. Here we create three buttons one to turn on the light, one to turn off the light and another to close the pop up, as we can see we send information to the group address 1/1/1 which is the one corresponding to our ceiling light that will turn on with 1 and it will go out with 0. The last button "Close" calls the script "closeForm" that is responsible for hiding the popup.

```html
<div class="w3-container" id="lighting"
style="margin-top:75px">
    <h1 class="w3-xxxlarge
    w3-text-red"><b>Iluminación</b></h1>
    <hr style="width:50px;border:5px solid red"
    class="w3-round">
    <button class="btn"
    onclick="openForm('lv_lso_popup')"><i class="fa
    fa-lightbulb-o"></i> Roof</button><br>
    <div class="form-popup" id="lv_lso_popup">
```

```
<form method="POST" class="form-container">
    <center><h2 class="w3-xxlarge
    w3-text-red"><b>Roof</b></h2></center>
    <button name="action" class="btn"
    value="knx-1/1/1-bin-1"><i class="fa
    fa-lightbulb-o"></i> Turn On</button><br>
    <button name="action" class="btn"
    value="knx-1/1/1-bin-0"><i class="fa
    fa-lightbulb-o"></i> Turn Off</button><br>
    <button type="button" class="btn cancel"
    onclick="closeForm('lv_lso_popup')">
    Close</button>
  </form>
</div>
```

Now we see the two scripts that we are going to need to show and hide the pop up. Starting with the one to show the pop up, it takes the sent id and makes it visible through the style.display property. In this same way, closing the pop up hides it.

```
<script>
    function openForm(id_popup) {
        var id = id_popup;
        document.getElementById(id).style.display =
        "block";}
    function closeForm(id_popup) {
        var id = id_popup;
        document.getElementById(id).style.display =
        "none";}
</script>
```

You can test the rest of the template, modifying to learn by doing it.

Monitor template (friki.html):

This template is interesting because here we use websockets to pass information in real time efficiently. So we will start with that.

For this we need to make a script that receives the information that the server sends us through the Flask application. For this we first define the socket through which we will receive the information and the function in which the information "my_response" comes from, as well as the name of the message. The information that the server sends us comes as a string so we have to make it a JavaScript object, we do this for the data and for the instructions. And finally what we do is to an element with a specific id we assign in its text field each of the records of both the data and the instructions.

```html
<script type="text/javascript" charset="utf-8">
    $(document).ready(function() {

        var socket = io.connect(location.protocol +
        '//' + document.domain + ':' +
        location.port);
        socket.on('my_response', function(msg) {

        var sys = String(msg.sdata);
        var system = JSON.parse(sys);
        var tel = String(msg.data);
        var telegrams = JSON.parse(tel);

        //Telegrams
        try{
          $('#idd0').text(
telegrams.telegrams[0].idd).html();
          $('#dated0').text(
telegrams.telegrams[0].dated).html();
```

```
            $('#cond0').text(
telegrams.telegrams[0].controllerd).html();
            $('#devd0').text(
telegrams.telegrams[0].deviced).html();
            $('#fund0').text(
telegrams.telegrams[0].functiond).html();
            $('#vald0').text(
telegrams.telegrams[0].valued).html();
            }
        catch(err){var e=err;}
        //SYSTEM_LOG
        try{
            $('#ids0').text(
system.logs[0].ids).html();
            $('#dates0').text(
system.logs[0].dates).html();
            $('#cons0').text(
system.logs[0].controllers).html();
            $('#devs0').text(
system.logs[0].devices).html();
            $('#funs0').text(
system.logs[0].functions).html();
            $('#vals0').text(
system.logs[0].values).html();
            $('#stas0').text(
system.logs[0].states).html();
            }
        catch(err){var e=err;}
            });
        });
    </script>
```

Identifiers type "# idd0", "dated0", "cond0", etc. They are cells of two tables created within the template in question.

And just like that we have the KNX bus information on the

server and then on the client in real time. This can be a bit overwhelming and the truth is that to understand just looking at these pages is not enough, you will have to take the codes from the repository and work with them until you understand them and create yours of course improved. So you will not only see that it is possible to create technology but you will feel it in your blood and I assure you that you will see things differently. This will require work on your part but I assure you that the satisfaction of seeing your own KNX server running will be incredible. If you feel stuck at any time do not hesitate to ask me for a hand and I will gladly guide you in the process of understanding the codes and putting your own KNX server to work.

If you want to accept a challenge, I encourage you to modify the templates with what you have learned so far and with the tools we have acquired so that on the home page the color of the light button changes with respect to the state of the light. You dare?.

Remember that if someone has already done something then you can also do it and if on the contrary nobody has done it then you should be the first to do it.

V

TRENDS AND PROJECTIONS

The market is changing and technology is advancing exponentially, that is why this last part of the book is of vital importance since it envisions what will come in the coming years, giving you the tools to anticipate your time.

"Innovation is what distinguishes one leader from the others" Steve Jobs

31

Training

The first thing that has to change and will surely be training in KNX. It is easy to see, see in your city how many KNX partners there are and of those accounts how many are in the capacity to successfully operate a KNX project. Now look at how many of them actually recovered the investment made in their training. With these results you will agree with me that something is not working in the correct way.

It is clear that there are very good teachers in KNX training and people with tremendous experience in the area, so why does it seem that the system is not working? Because it is precisely the system and the bureaucracy that is not really working, education is a business and that is fine, what is wrong is that it is not a clear business where the parties enter knowing that they are going, leave satisfied and after some time they will recover their investment by far, whatever it may be. The diploma is very nice but the truth is paper and that serves only to feed your own ego, perhaps to fulfill some requirements or impress someone but not more. What really matters is what you know and the person you become after education.

They say that homeschooling is the most important and to this day I don't know the first person to get a diploma for it. It is absolutely irrelevant that, what is crucially important is the person you become and that is your tools for the rest of your life. Well, here it should be the same, you should be a different person with very clear tools to put into practice in real life and improve your quality of life. Of course if you also have a diploma, a nice title and a medal even better !!.

The big problem with KNX training is that students arrive with a painted world of illusions full of prosperity and abundance where they will invest little amount of money. When they enter the courses they realize that most of what they see will serve them absolutely nothing more than to pass the exam and obtain their diploma. In some cases they see teachers at the level where there are no mistakes and everything seems easy, but it is not necessary to think since they paint example situations and how to get out of them without problems. After several days of playing with training suitcases and reading dozens of pages at night, they find an exam where they pass a diploma and go out with it to face the world. When the time comes for a project they realize that what they suspected during the training about the uselessness of most of the content was real and they crash into a large wall, where they make mistakes for the first time that during their games with the suitcases did not happen and therefore they do not have the remotest idea of how to solve them.

Even worse, the vast majority of those who get their diplomas never get to do a project ending in frustration and an absolutely zero return on investment.

This should not be so, every day more and more partners come out increasing the supply irresponsibly knowing that the demand, at least for now, is not increasing in the same way. This is detrimental to both individuals and KNX as a global organization, after all an organization is its people and partners are part of that organization so its success in the sector will be directly associated with that of KNX.

This is normal, it is a stage of evolution in training and it is very difficult for KNX as an organization to take control of all this but the bureaucracy and the system in general will have to change.

Training should focus on creating a change in the lives of students who want to be partners, but a real and tangible change. This includes improving their ability to reason, to create, their residence and ultimately their quality of life associated with their income. To achieve this, you must start by filtering the information as I have tried to do in this book and provide what will really be more useful and applicable in the daily life of a student than to become a professional in the automation of homes and buildings. Surely many technical foundations are useful only for those who make developments and this for those who do not do it is just noise, then a course focused on development and another will be created. Developers and integrators are different profiles so their education must be different with what is really useful for each of them, a healthy ecosystem must have diversity and different types of people.

Once the theoretical information has been refined and focused, it must be complemented with practical information to accelerate the learning curve of the students. Fortunately, there are great professionals at KNX with much to share with those new to the sector. Thus, new refined and fully applicable information will be generated for each student profile, complemented with access to virtual simulators that allow students to practice autonomously before each practice session.

During practical training, monotonous and trivial practices should be abolished, these contribute nothing more than what the student could do at home with a simulator. The days of face-to-face practice should be full of challenges and problems that the students will try to solve and will rely on the tutors who will teach them to generate logical thinking structures in the KNX system, so that when problems arise outside the classroom they have full capacity to solve them looking for the information they need to be the case.

Finally, the supply and demand of professionals must be controlled through the association of each country to avoid saturating the market, damaging the sector and its level, both technically and economically and commercially.

32

Market

The market for home and building automation will be divided into three main groups. And you will have to project which of the three groups you will focus on your business model, let's see.

The first group is made up of hobbyists, automation and home automation enthusiasts who focus on using "Do It Yourself" equipment. That is, easy-to-use equipment and the well-known "Gadget". In this group will be engineers or people who like technology, who see devices as toys for their entertainment and fun. For them, the intermediaries between the factories of these types of products and they are absolutely censers, because this group likes to buy the equipment and learn to use and integrate it by themselves. It should be noted that in this category the projects as such are non-existent, focused on products that fulfill a specific function and are integrated at least partially with the cell phone or other equipment. If you want to enter this market, basically the way is to develop products or services, that is, you will have to become a manufacturer. For example, the smart light bulbs that connect to the cell phone and that's it, that's a DIY product, you don't need an integrator or anything like that to put it to work and fulfill its specific function.

The second group is people or institutions interested in professional automation projects. Here the integrators are absolutely necessary and the products used here are of higher technical specifications and require specific knowledge to be able to start them. Here is KNX for example, unlike the previous group these devices are not usually sold directly to end users by manufacturers because it is not as simple as plugging in and enjoying. For example, a KNX button, by itself is useless as well as there are a thousand options, it is necessary to have a whole system and make specifications and then select the equipment and in the end have a professional automation project. In this group, the integration and interoperability of the systems will be fundamental, more and more open protocols will gain strength and in the future, due to the number of devices and brands, it will be almost an obligation to use an open protocol or to have the option of joining one.

The third group will be companies or institutions focused on the elderly. This group will be very interesting because its main objective will be the health care of the elderly through the automation of homes and buildings. For them, the protocols will be irrelevant and the main objective will be to monitor people and their activities to associate this with the health system with medical personnel who, based on the information obtained, will be able to identify if people are at risk or have alterations in their daily behavior. For this, artificial intelligence is a fundamental pillar so let's go see it.

33

Artificial intelligence

Artificial intelligence has been one of the interesting trends of recent years and will continue to be so in the coming years. The theory is well developed and there are numerous tools to work with artificial intelligence. What will come will be innovation, applying it to different sectors. In the sector of home and building automation, there is significant research in the application of artificial intelligence in the recognition and prediction of daily activities within an intelligent environment.

Daily activities are essential in health care, especially for the elderly. This is because many illnesses can be identified early due to alterations in daily activities or even risky situations can be anticipated, for example, imagine a person who stops eating, starts taking their medicine late, or starts getting up in the middle of the night. If a person can live at home or in an institution with the infrastructure to obtain this information and analyze it, their quality of life will increase very significantly.

In a smart home there are many variables that are directly correlated with our behavior and through them you can classify the activities that are being carried out, and then predict them. This is where artificial intelligence comes in by taking all of these variables and doing a classification of activities, this is the first approach and the second approach is to take the same variables to predict future activities in anticipation of the person even thus helping people who lose their memory. This is very interesting because there are various research and academic projects with these two approaches, only that there is a gap between industry and academia. In this area, research has had and will continue to have three main approaches.

The first is based on sensors carried by people. These sensors are typically used by people. The strategic location of these can provide information on the movements of different parts of people's bodies and develop models of activities. These sensors include inertial units, sensors embedded in smartphones, radio frequency identification sensors and RFID tags. In this approach, two main problems are analyzed, the first is the recognition of activities as such and the second is the location of the person, which improves the recognition of activities.

The second is based on sensors located in the environment. This is the broadest approach and with the greatest variety of sensors, here are all those sensors that are carried by the person or cameras or microphones. These sensors are commonly embedded in the environment, that is, in the automated house, among which we can highlight movement sensors, magnetic sensors for opening doors, temperature sensors, pressure sensors, air quality sensors, weather stations and current sensors for electrical equipment. Investigations of this approach often use public databases of sensor data in test houses, such as the CASAS project at Washington State University.

The third is based on cameras and microphones. This approach is usually the most controversial since it uses cameras and microphones inside the home, which generates great skepticism related to privacy. In this approach, image processing, video and microphones are used to determine the activity that the person is executing. Normally it requires a greater computing capacity and uses machine learning techniques, but it has a great advantage and is the ability to identify different users doing different activities, which in the previous approaches generates a totally different problem, which is the handling of data in multi-user environments. However, its implementation potential is very limited due to the high sensitivity in privacy within an automated house.

Identifying the activities of the person in the home has great advantages since it allows extrapolating behavior patterns and analyzing the history of activities carried out, it is possible to detect diseases related to behavior modification, which when diagnosed early allow their correct treatment. Once the data is obtained, the labeling of the activities is carried out, which is mainly done later and with human analysis. It should be noted that this is also the first step in predicting activities that we will discuss below.

Once the activities that the person is executing are identified, the next step is to seek to predict them in order to anticipate them and take preventive actions. This in order to achieve energy efficiency, facilitate people's lives or prevent critical situations. However, the prediction of activities of daily living has been less developed than the recognition of them. Most activity prediction publications and projects start from an open database with a set of activities that are the product of data taken in automated test houses or apartments. The main reason for this is the work density that implies the implementation of a smart home and the complexity of recognizing the activities themselves, these are two complete problems that are usually solved to look for the prediction of activities. Different methods based on artificial intelligence techniques are used for this prediction.

We will start with the Bayesian networks. In the probabilistic graphical models (PGMs) the variables are represented as nodes and the statistical dependence between the respective random variables as arcs. Therefore, PGMs provide a compact way of representing the joint probability distribution. These directed graphical models are known as Bayesian networks. These networks are used to predict activities, taking them as variables and adding descriptive state variables of the activity in question. Constraint, search, and score-based models are also used. The sequences of activities with Bayesian network models also show good results.

There are also the hidden Markov models. To predict the activities that the person will take home, it is necessary to predict behavior patterns, which are affected by environmental variables that are sometimes hidden from the system. This change of hidden states generates a change in the observed states, in this case the activities carried out by the user. In order to approach this approach, hidden Markov models (HMM) are implemented, which have been widely used in predicting user behavior. An HMM is a dual scholastic model described by a finite number of hidden states, a finite number of discrete observation symbols, a state transition probability matrix, an emission probability matrix, and an initial state probability distribution. Because each user has different behavior patterns, a custom model is usually generated for each one, using iterative algorithms such as Baum-Welch to find the HMM that fits the observations.

Machine learning is also used. Machine learning techniques are mainly useful where patterns are fixed with "oine" learning, on the other hand when patterns are dynamic and changing with respect to the environment or the behavior of users, "online" learning is used. Among the models used are models based on MDP with q-learning, deep learning, bootstrapping, support vector machines and grouping-based classification.

And of course there are neural networks. Neural networks are widely used for their versatility and their potential to solve problems where it is not feasible to find an exact solution due to the number of unknown factors, however their use in predicting activities within a smart home needs to be further explored. Among its main applications is HVAC advance management.

There are numerous models and techniques for recognizing activities for the subsequent prediction of activities or actions that the user will carry out within the smart home. However, two problems stand out mainly in order to bring this approach to a commercial implementation that impacts the lives of millions of people. The first is the separation between recognition of activities for subsequent labeling to subsequently make the prediction, this is a problem because it requires constant intervention to the system and makes it inflexible to changes in patterns or infrastructure of the home. The second problem is the little flexibility of the models and the system of both recognition and prediction of the change of input variables, which is fundamental for a real commercial implementation.

To make all this possible, it is essential to obtain the different data in a more efficient way and to interconnect the different computers and this is where the Internet of Things comes into play.

<h1 style="text-align:center">34</h1>

Internet of things

The Internet of Things is not really new, it has been on the market for several years and home and building automation systems can be connected to the Internet to manage the system remotely. However, this is not the internet of things, just internet connectivity. The internet of things takes place when everyday objects connect to the network to extend its functionality by interacting with other devices.

The IoT is currently very focused on very focused automation, that is, on very specific tasks and most at the level of enthusiasts or DIY. They are devices with direct internet connection and that can be quickly started to fulfill their particular function, the interesting thing begins when we want these devices to interact with each other or we want to apply them in professional automation. Here there is a small gap that will gradually close.

We are going to analyze several things in the current internet of things. The first is that the system is absolutely dependent on the internet, which is not so good because if there is no internet it just doesn't work. So here we think of radio frequency as a direct solution, however expanding the panorama we can see that the solution is devices connected to IP instead of RF. Imagine a city where each house is full of RF devices, without a doubt the range of many will cross and it is much more difficult to identify and differentiate them all. On the other hand, the devices connected by means of IP would be encapsulated within the local network of each house, which makes management much more efficient. Let's also think about security, it is much easier to create a secure local network that even detects intrusion or perhaps obstructions than a secure RF network, an RF network in principle is easily lockable. So the creation of secure local internet networks will have great importance in the near future applied to the automation of homes and buildings.

The second relevant aspect with these IoT devices is the integration with professional automation systems, you could usually think of a Gateway. The problem here is that a breakpoint is generated in the integration as there are numerous devices integrated by a single gateway. To solve this, home and building automation product networks will have to be native IoT, just as twisted pair is native to all products and different mechanisms have been generated to ensure their reliability, future networks will have to focus on network interconnection. This will generate a paradigm shift since it will not be the IoT devices that are integrated into home and building automation systems with conventional devices, but it will be currently conventional devices that will be integrated into home and building automation systems. IoT.

To achieve this, it is important that the variables have a context instead of just a value. This can either be in a central controller in charge of generating these semantics or they can be included in the original variable. Generating autonomy in the information, that is to say, when seeing a telegram that travels through the network, it will be possible to know if it comes from a refrigerator, a light, a blender, a video door entry unit or another device unlike current technology where only one knows that it comes of certain addresses without having more information about those addresses to which they belong or in which context they are found.

Currently this can be done by software in a central controller and generate these complete models where the variables are condensed with their respective context but this in turn generates a drawback, which all depends on the central controller and if this fails the system will be as conventional as 15 years ago. This is why these models must be generated directly in the equipment in a standardized and efficient way, so that in case of failure of the central system, the BMS or the display system, the automation system does not suddenly fall in a leap into the past.

35

Augmented reality

S mart home touchscreens are going to disappear and become obsolete. We will still need a couple of years to get to that point but it will be an inevitable reality.

We will start with the closest thing, the next milestone in this area. Every day the televisions have more powerful electronics and their operating systems become more robust, so what sense does a 7-inch screen have in the middle of your room if right in front you have a 40-inch one with very high resolution, much more powerful both hardware and software and also interactive. Yes, I know, you will tell me that it is more practical to have a small screen more at hand, well you already have a smaller, more comfortable, more intuitive and also more powerful ... your cell phone.

At the beginning this idea may seem shocking and more if you are the manufacturer of the screens or if you sell them, but if that is your case, it is time to start generating that migration and challenge the status quo. The tools are already on the table and today what I am saying could come true. Smart TVs have mostly embedded Linux processors, they even have an operating system that runs on the same basis as current KNX integrations. These smart TVs have applications developed for their own operating system with frameworks that currently have KNX integration models, this is very powerful. It means that today you can make an application that controls your entire KNX system from your TV without gateways or anything like that, directly. This is going to generate a great change in the way of thinking about automation projects and even the manufacture of smart televisions.

As soon as the big TV manufacturing companies realize this opportunity and the potential that their own TV, with a small investment in the development of the necessary software, can give the added value of controlling the entire KNX installation being user configurable. it will become a trend. All manufacturers will want to offer the same option to avoid becoming obsolete in the market and this in turn will drive the market to manage the systems from their own television with all the advantages that this entails. Among them, cost efficiency, greater interactivity, high resolution, integration with audio and video, among others.

Ultimately this will be extended to augmented reality, projecting virtual screens and information in current reality. Thus, users interact naturally with the system, which will no longer be limited to a screen, but instead will be able to project interactive virtual objects on any environment in the home. This will lead to the integration of automation systems and the extension of people's work or entertainment environments.

And the best news is that although this is not yet possible and the technology of the moment limits us a little, there are bases that allow the integration of KNX to be guided to augmented reality, even if initially.

About the author

Miguel Ángel Rodríguez Segura is an electronic engineer graduated with honors from the Pontificia Universidad Javeriana de Colombia, specialist in Strategic Management and Innovation at Copenhagen Business School, specialist in Digital Marketing at University of Illinois, specialist in Deep learning at deeplearning.ai , specialist in Full-Stack Web Development with REACT from the Hong Kong University of Science and Technology, Master in Electronic Engineering by research from the National University of Colombia and expert in home and building automation. He is a KNX tutor and is certified in different protocols for home and building automation as well as industrial automation. He is a tech based entrepreneur.

During his undergraduate thesis, he worked in nanotechnology and biochip creation, where he was awarded and nominated for the best thesis in the country. Later, he was a practitioner in a multinational, manufacturer of KNX equipment, among many other things, where he went into control and automation, then went on to work in the banking sector in software automation for a short time, and from there he chose to create and lead departments. automation in the private sector.

Later he decided to undertake in the sector of home and

building automation, being also a lecturer at the Colombian association of engineers ACIEM and consultant in building automation and home automation. He has taught different specialized courses to integrating companies, engineers, training centers and manufacturers on deep automation topics as well as KNX certifications.

He has trained with the best in the world and has participated in large projects. Also as a consultant and auditor, he has seen and solved many problems in the different automation sectors. He has always been innovative and risky throughout his career venturing into new terrain, which has allowed him to learn from it and improve at an accelerated rate. In addition, he has had the fortune to share with people with a lot of experience and even with more training than him, receiving both theoretical and practical knowledge that is reflected in this book.

Acknowledgements

This book has been made possible by many people who have contributed to my academic, professional and personal training. We are a product of both ourselves and the people who have surrounded us during our lives and fortunately I have had luck in this.

Special thanks to Linda, Clara and Juan Carlos for supporting me at the beginning of my professional career. I am fortunate to have found incredible bosses from whom I learned a lot and whom I deeply admire. To Nelson thank you very much for being an example to follow.

Thank you to the people who have given me the great privilege of participating in your automation projects, it has been incredible.

To the KNX International Association, of course, thanks for such a good system and for the incredible work they do every day, my respect and admiration for them. Jesus thank you for putting your soul into KNX and helping so many people, it is very nice to see such capable and kind people in important positions, keep up the great work.

To my colleagues for their feedback and their teachings.

To Julieth for her support with the translation fo the book.

To Jose Maria and Julio for being my formal trainers at KNX, I have been fortunate to learn from the best and I am very happy about it.

To Jose Angel for his support and advice, for shared knowledge and for being an example to follow. Total admiration and my absolute respects.

To my partners who have helped shape my career and my businesses, with whom we had a hard time and mature ones. Literally thanks for joining me on the adventures that have taught me so much.

To my family Miguel, Nubia and Giselle for their incredible support and inspiration, without them there literally would not exist and therefore neither would this book. My father is a brilliant teacher of science and life with incredible momentum and an incomparable heart, what a great example I have had. My mother, a teacher who has taught me a lot of wisdom apart from physics, it is difficult to find these people who transmit serenity and find the best way to do things, I have always had them by my side. My sister, an incredible talent that is developing, has always inspired me to be better and she is a great engine in my life, thank you very much.

And of course to you! Yes to you reader for giving you the opportunity to learn KNX in a different way or complement your training, giving me the pleasure of contributing to it. Hopefully one day we will have the opportunity to meet, a big hug and my best wishes on your next adventures.

Always remember that boats were not made to anchor in port forever, they are made to sail so good wind and good sea!

With affection, Miguel Angel.

www.ingramcontent.com/pod-product-compliance
Lightning Source LLC
Chambersburg PA
CBHW031039160726
47991CB00005B/1943